Convergence of Intelligent Data Acquisition and Advanced Computing Systems

Convergence of Intelligent Data Acquisition and Advanced Computing Systems

Editors

Grigore Stamatescu
Anatoliy Sachenko
Dan Popescu

MDPI • Basel • Beijing • Wuhan • Barcelona • Belgrade • Manchester • Tokyo • Cluj • Tianjin

Editors
Grigore Stamatescu
Automation and Industrial Informatics
University Politehnica of Bucharest
Bucharest
Romania

Anatoliy Sachenko
Information Computer Systems and Control
West Ukrainian National University
Ternopil
Ukraine

Dan Popescu
Automation and Industrial Informatics
University Politehnica of Bucharest
Bucharest
Romania

Editorial Office
MDPI
St. Alban-Anlage 66
4052 Basel, Switzerland

This is a reprint of articles from the Special Issue published online in the open access journal *Sensors* (ISSN 1424-8220) (available at: www.mdpi.com/journal/sensors/special_issues/IDAACS2019).

For citation purposes, cite each article independently as indicated on the article page online and as indicated below:

LastName, A.A.; LastName, B.B.; LastName, C.C. Article Title. *Journal Name* **Year**, *Volume Number*, Page Range.

ISBN 978-3-0365-1656-1 (Hbk)
ISBN 978-3-0365-1655-4 (PDF)

Contents

About the Editors vii

Preface to "Convergence of Intelligent Data Acquisition and Advanced Computing Systems" ix

Jakob Pfeiffer, Xuyi Wu and Ahmed Ayadi
Evaluation of Three Different Approaches for Automated Time Delay Estimation for Distributed Sensor Systems of Electric Vehicles
Reprinted from: *Sensors* **2020**, *20*, 351, doi:10.3390/s20020351 1

Dan Popescu, Florin Stoican, Grigore Stamatescu, Loretta Ichim and Cristian Dragana
Advanced UAV–WSN System for Intelligent Monitoring in Precision Agriculture
Reprinted from: *Sensors* **2020**, *20*, 817, doi:10.3390/s20030817 19

Georgios Karalekas, Stavros Vologiannidis and John Kalomiros
EUROPA: A Case Study for Teaching Sensors, Data Acquisition and Robotics via a ROS-Based Educational Robot
Reprinted from: *Sensors* **2020**, *20*, 2469, doi:10.3390/s20092469 45

Roozbeh Sadeghian Broujeny, Kurosh Madani, Abdennasser Chebira, Veronique Amarger and Laurent Hurtard
Data-Driven Living Spaces' Heating Dynamics Modeling in Smart Buildings using Machine Learning-Based Identification
Reprinted from: *Sensors* **2020**, *20*, 1071, doi:10.3390/s20041071 63

Rytis Augustauskas and Arūnas Lipnickas
Improved Pixel-Level Pavement-Defect Segmentation Using a Deep Autoencoder
Reprinted from: *Sensors* **2020**, *20*, 2557, doi:10.3390/s20092557 79

Muhammad Jawad, Muhammad Bilal Qureshi, Sahibzada Muhammad Ali, Noman Shabbir, Muhammad Usman Shahid Khan, Afnan Aloraini and Raheel Nawaz
A Cost-Effective Electric Vehicle Intelligent Charge Scheduling Method for Commercial Smart Parking Lots Using a Simplified Convex Relaxation Technique
Reprinted from: *Sensors* **2020**, *20*, 4842, doi:10.3390/s20174842 101

Ahmad M. Karim, Hilal Kaya, Mehmet Serdar Güzel, Mehmet R. Tolun, Fatih V. Çelebi and Alok Mishra
A Novel Framework Using Deep Auto-Encoders Based Linear Model for Data Classification
Reprinted from: *Sensors* **2020**, *20*, 6378, doi:10.3390/s20216378 121

Vahid Tavakkoli, Kabeh Mohsenzadegan, Jean Chamberlain Chedjou and Kyandoghere Kyamakya
Contribution to Speeding-Up the Solving of Nonlinear Ordinary Differential Equations on Parallel/Multi-Core Platforms for Sensing Systems
Reprinted from: *Sensors* **2020**, *20*, 6130, doi:10.3390/s20216130 143

Oleksandr Drozd, Grzegorz Nowakowski, Anatoliy Sachenko, Viktor Antoniuk, Volodymyr Kochan and Myroslav Drozd
Power-Oriented Monitoring of Clock Signals in FPGA Systems for Critical Application
Reprinted from: *Sensors* **2021**, *21*, 792, doi:10.3390/s21030792 161

About the Editors

Grigore Stamatescu

Grigore Stamatescu graduated from the University Politehnica of Bucharest (UPB) in 2009 and holds a PhD degree (2012) from the same university. He is currently an Associate Professor (Habil. 2019) with the Department of Automation and Industrial Informatics, Faculty of Automatic Control and Computers, UPB. His research interests include networked embedded sensing, the Internet of Things and distributed information processing in the industry, the built environment, and smart city applications. His recent research focused on statistical learning methods for load forecasting and anomaly detection in building energy traces, and data-driven modelling of large-scale manufacturing systems, with a focus on energy efficiency. His research has been published in over 130 articles. Dr. Stamatescu was a Fulbright Visiting Scholar in 2015–2016 at the University California, Merced, and a JESH Scholar of the Austrian Academy of Sciences in 2019. He is a Senior Member of IEEE.

Anatoliy Sachenko

Anatoliy Sachenko is a Professor and the Head of the Department of Information Computing Systems and Control, and a research advisor of the Research Institute for Intelligent Computer Systems, West Ukrainian National University. He earned his B.Eng. degree in Electrical Engineering at L'viv Polytechnic Institute in 1968, his PhD degree in Electrical Engineering at L'viv Physics and Mechanics Institute in 1978, his Doctor of Technical Sciences Degree in Electrical and Computer Engineering at Leningrad Electrotechnic Institute in 1988. Since 1991, he has been an Honored Inventor of Ukraine, and since 1993, he has been an IEEE Senior Member. His main areas of research interest are the implementation of artificial neural networks, distributed systems and networks, parallel computing, and intelligent controllers for automated and robotics systems. He has published over 450 papers in the areas listed above.

Dan Popescu

Dan Popescu was born in Bucharest, Romania, in 1950. He received his BS and MS equivalent (five-year engineering) degrees in Automatic Control and Computers from the University Politehnica of Bucharest in 1974; his BS and MS equivalent degrees (five-years) in mathematics from the University of Bucharest, Romania; and his PhD degree in automation and remote control from the University Politehnica of Bucharest in 1987. Since 2003, he has been a Full Professor with the Faculty of Automatic Control and Computers; a PhD supervisor and the Head of the Laboratory of Innovative Products and Processes for Increasing Quality of Life PRECIS Center, since 2016; and the Vice-Dean of Research Activity with the University Politehnica of Bucharest from 2012 to 2016. His main research interests include image processing and interpretation, neural networks, wireless sensor networks, control systems in industrial and robotic applications, and environmental monitoring.

Preface to "Convergence of Intelligent Data Acquisition and Advanced Computing Systems"

This preface briefly outlines the objectives of the Special Issue on "Convergence of Intelligent Data Acquisition and Advanced Computing Systems" published between September 2019 and September 2020 in the journal *Sensors*. This Special Issue welcomed submissions as extended versions of conference articles published in the proceedings of the "10th IEEE International Conference on Intelligent Data Acquisition and Advanced Computing Systems: Technology and Applications (IDAACS'2019)", allowing the authors to expand on their theoretical and experimental contributions. These were complemented with external submissions from interested researchers working in this timely area. We highlight the main contributions of the published articles, grouped by key topics: intelligent data acquisition, learning methods and optimization for data processing, and advanced computing systems to support data processing.

As modern data-acquisition solutions increasingly integrate on-board computing in the case of intelligent sensors, in-network computing, or in the case of distributed systems and sensor networks, a need to provide a joint view of these previously divergent topics has been observed. We define the convergence of intelligent data acquisition and advanced computing systems at the interface of theory and applications for industrial manufacturing, scientific computing, precision agriculture, and energy systems such as smart building and smart city systems. This book bridges specific topics for instrumentation such as accuracy, sampling, time synchronization, sensor selection and calibration with algorithm design, statistical and machine learning, computational efficiency, and reconfigurable computing that supports conventional engineering tasks. Computing-as-a-service is another relevant approach that can be leveraged to improve measurements and instrumentation in the future.

The potential topics of this Special Issue that were initially defined were mapped onto the core areas and special streams of the IDAACS'2019 conference and included but were not limited to advanced instrumentation and data-acquisition systems; advanced mathematical methods for data acquisition and computing systems; computational intelligence for instrumentation and data-acquisition systems; data analysis and modeling; intelligent distributed systems and remote control; intelligent information systems, data mining, and ontology; Internet of Things; pattern recognition, digital image, and signal processing; intelligent instrumentation and data acquisition systems in advanced manufacturing for Industry 4; intelligent robotics and sensors; machine learning; and application for smart buildings and smart cities.

This Special Issue received 25 submissions, and after a thorough and competitive review process, only nine articles were published. The accepted articles were co-authored by 41 authors in total from 14 countries, with good geographical diversity. Performing an overview of the accepted articles, we grouped them into three main categories focused on data acquisition, modern data-processing algorithms, and computing architectures that support data processing.

Intelligent data acquisiton and data collection platforms: this section highlights three articles that focus on data acqusition from sensors and sensor platforms in electrical vehicles, precision agriculture, and educational robotics applications.

In [1], the authors provided an experimental evaluation of time delay estimation methods for current measurements in electrical vehicle powertrains. These include linear regression (LR), variance minimization (VM), and adaptive filters (AF). The methods were benchmarked in terms of Root Mean Squared Error (RMSE) and average run-time on data collected from realistic driving profiles. Given its efficiency, the VM method was recommended for this application considering noise resistance and computational efficiency. This study is highly relevant as it also considers the computational constraints of distributed electronic control units (ECU) in modern automobiles.

Ref. [2] presented a system that combines scalar, ground-level measurements from sensor networks with aerial robotic platforms (UAVs) as a data-collection and communication-relay infrastructure. The main innovation lies in the network data aggregation by consensus in order to reduce the need for data transmissions. The flight path of the UAV was optimized using spline functions in order to maximize the flight time and data gathering in a given area from the cluster heads. A symbolic aggregation approximation (SAX) method encoded the raw sensor measurements into text strings for each of the monitored parameters in the precision agriculture application. The results quantify both the data reduction as well as the UAV performance improvement using the trajectory control scheme.

A sensor-intensive robotic platform for education was introduced by [3]. A thorough review of existing educational robotic platforms was carried out alongside the argument for the Robotic Operating System (ROS) as a framework of choice for the underlying implementation and new developments. The system was built around a Raspberry Pi-embedded development board with dedicated function specific sensors such as ultrasonic sensors, wheel encoders, LIDAR, and a camera. It can be controlled over a WiFi communication interface using a PC or a mobile phone. Its performance was experimentally validated through a series of tests as well as in qualitative studies using various groups of students.

Learning from data and optimization: this section focuses on four articles that cover statistical and machine learning and optimization techniques that make use of collected sensor data together with advanced computing algorithms for implementation.

In [4], a detailed study was provided for smart building system identification combining both classical, such as nonlinear autoregressive (NARX) models, and data-driven machine learning-based approaches, such as multi-layer perceptrons (MLP). The goal was to achieve an accurate model of building thermal dynamics for control using collected data under various conditions. The data were collected through an existing building automation system (BAS) infrastructure with wireless components for sensing, room temperature sensors, and controlled motor valves for the heating elements. The study was carried out on a real building at the University Paris-Est Creteil (UPEC). The main results showed good performance in terms of Mean Squared Error (MSE) and Mean Absolute Error (MAE), less than 0.2C compared to the ground truth.

Reference [5] introduced an improved deep learning method for the evaluation and classification of road quality based on the U-Net deep autoencoder. The main innovation stemmed from the addition of residual connections, atrous spatial pyramid pooling, and attention gates to increase performance. An evaluation was performed on multiple reference benchmarking datasets: CrackForest, Crack500, and GAPs384. The Dice score was used to compare various architectures and parametrization options of the deep learning architectures with a robust improvement in the proposed ResU-NET + ASPP + AG network over the baseline specfication. Testing was performed on dedicated graphical processing units on each dataset, while mixed dataset training did not yield consistent results.

A mixed integer linear programming optimization problem for smart parking EV charging was formulated by [6]. The goals were two-fold: to maximize the parking lot revenue by accommodating charging EVs as efficiently as possible and by minimizing the cost of power consumption through participation in the utility-level demand response (DR) program. The study was validated in a simulation using predefined schedules and model EV charging characteristics. A simplified convex relaxation technique was introduced to ensure the feasibility of the optimization problem. The solution was compared against a standard variable charging power approach and showed consistent improvements in terms of power consumption cost and percentage savings.

The authors of [7] proposed a new scheme for data classification by combining sparse auto-encoders (SAE) with data postprocessing using a nature-inspired particle swarm optimization (PSO) algorithm. The postprocessing layer improved the classification performance of the deep neural network through a parameter optimized linear model. The labeled datasets used for practical evaluation stemmed from the medical field and included epileptic seizures, SPECTF, and cardiac arrhythmias while experimenting with multiple parameters of both the neural network and the PSO algorithm. The adjustment layer improved the performance of the models, as illustrated through other documented studies from the literature, achieving for example a 99.27% accuracy on the cardiac arrhythmia dataset.

Advanced computing systems for data processing: this section includes two articles that present improvements to data processing through optimized architectures for solving differential equations and reconfigurable computing with FPGAs.

Reference [8] approached the problem of increasing the performance when solving ordinary differential equations (ODE) on multi-core embedded systems, which can describe the system model of certain physical phenomena. The authors introduced an adaptive algorithm, PAMCL, based on the Adams–Moulton and Parareal methods and provided a comparison with existing approaches. The implementation-wise OpenCL platform was used with optimized solvers for both CPU and GPU systems. Quantitative results were reported, which include the CPU run time, GPU speed-up, and the memory footprints of the reference algorithms. This method showed good results and achieved full convergence to the exact solutions. A potential extension of the PAMCL method for partial derivative equations was described.

Power-oriented monitoring of clock signals in FPGA systems was described by [9]. The argumentation of the work lays out the need to reduce the power consumption and checkability of reconfigurable computing platforms. The study included two types of power-monitoring: the detection of synchronization failures, and the dissipation of power using temperature and current sensors. The experiments were carried out on typical computing tasks, e.g., digital filter implementations, using standardized tools for monitoring and data collection. The thermal and power dissipation data were associated with fault conditions in the synchronization. Such improvements to the evaluation of FPGA systems are highly relevant for critical and highly reliable applications.

References:

1. Pfeiffer, J.; Wu, X.; Ayadi, A. Evaluation of Three Different Approaches for Automated Time Delay Estimation for Distributed Sensor Systems of Electric Vehicles. Sensors 2020, 20, 351. [Google Scholar] [CrossRef] [PubMed]

2. Popescu, D.; Stoican, F.; Stamatescu, G.; Ichim, L.; Dragana, C. Advanced UAV–WSN System for Intelligent Monitoring in Precision Agriculture. Sensors 2020, 20, 817. [Google Scholar] [CrossRef] [PubMed]

3. Karalekas, G.; Vologiannidis, S.; Kalomiros, J. EUROPA: A Case Study for Teaching Sensors, Data Acquisition and Robotics via a ROS-Based Educational Robot. Sensors 2020, 20, 2469. [Google

Scholar] [CrossRef] [PubMed]

4. Sadeghian Broujeny, R.; Madani, K.; Chebira, A.; Amarger, V.; Hurtard, L. Data-Driven Living Spaces'Heating Dynamics Modeling in Smart Buildings using Machine Learning-Based Identification. Sensors 2020, 20, 1071. [Google Scholar] [CrossRef] [PubMed]

5. Augustauskas, R.; Lipnickas, A. Improved Pixel-Level Pavement-Defect Segmentation Using a Deep Autoencoder. Sensors 2020, 20, 2557. [Google Scholar] [CrossRef] [PubMed]

6. Jawad, M.; Qureshi, M.B.; Ali, S.M.; Shabbir, N.; Khan, M.U.S.; Aloraini, A.; Nawaz, R. A Cost-Effective Electric Vehicle Intelligent Charge Scheduling Method for Commercial Smart Parking Lots Using a Simplified Convex Relaxation Technique. Sensors 2020, 20, 4842. [Google Scholar] [CrossRef] [PubMed]

7. Karim, A.M.; Kaya, H.; Güzel, M.S.; Tolun, M.R.; Çelebi, F.V.; Mishra, A. A Novel Framework Using Deep Auto-Encoders Based Linear Model for Data Classification. Sensors 2020, 20, 6378. [Google Scholar] [CrossRef] [PubMed]

8. Tavakkoli, V.; Mohsenzadegan, K.; Chedjou, J.C.; Kyamakya, K. Contribution to Speeding-Up the Solving of Nonlinear Ordinary Differential Equations on Parallel/Multi-Core Platforms for Sensing Systems. Sensors 2020, 20, 6130. [Google Scholar] [CrossRef] [PubMed]

9. Drozd, O.; Nowakowski, G.; Sachenko, A.; Antoniuk, V.; Kochan, V.; Drozd, M. Power-Oriented Monitoring of Clock Signals in FPGA Systems for Critical Application. Sensors 2021, 21, 792. [Google Scholar] [CrossRef] [PubMed]

Grigore Stamatescu, Anatoliy Sachenko, Dan Popescu
Editors

sensors

MDPI

Article

Evaluation of Three Different Approaches for Automated Time Delay Estimation for Distributed Sensor Systems of Electric Vehicles

Jakob Pfeiffer [1,2,*], Xuyi Wu [2] and Ahmed Ayadi [2]

1 BMW Group, Petuelring 130, 80788 Munich, Germany
2 Department of Electrical and Computer Engineering, Technical University of Munich, Arcisstr. 21, 80333 Munich, Germany; Xuyi.Wu@tum.de (X.W.); Ahmed.Ayadi@tum.de (A.A.)
* Correspondence: Jakob.J.Pfeiffer@bmwgroup.com

Received: 3 December 2019; Accepted: 26 December 2019; Published: 8 January 2020

Abstract: Deviations between High Voltage (HV) current measurements and the corresponding real values provoke serious problems in the power trains of Electric Vehicles (EVs). Examples for these problems have inaccurate performance coordinations and unnecessary power limitations during driving or charging. The main reason for the deviations are time delays. By correcting these delays with accurate Time Delay Estimation (TDE), our data shows that we can reduce the measurement deviations from 25% of the maximum current to below 5%. In this paper, we present three different approaches for TDE. We evaluate all approaches with real data from power trains of EVs. To enable an execution on automotive Electronic Control Units (ECUs), the focus of our evaluation lies not only on the accuracy of the TDE, but also on the computational efficiency. The proposed Linear Regression (LR) approach suffers even from small noise and offsets in the measurement data and is unsuited for our purpose. A better alternative is the Variance Minimization (VM) approach. It is not only more noise-resistant but also very efficient after the first execution. Another interesting approach are Adaptive Filters (AFs), introduced by Emadzadeh et al. Unfortunately, AFs do not reach the accuracy and efficiency of VM in our experiments. Thus, we recommend VM for TDE of HV current signals in the power train of EVs and present an additional optimization to enable its execution on ECUs.

Keywords: automotive; current; electric power train; electric vehicle; embedded systems; delay; detection; distributed systems; measurements; power train; sensor; signals; time delay estimation

1. Introduction

Political guidelines in various countries to decarbonize individual mobility led to an exponential growth of Electric Vehicles (EVs) in offers and sales. However, one obstacle for the success of EVs is the so-called range anxiety [1]. Customers are afraid that an EV is not able to provide the range they need for all of their journeys. To combat range anxiety and increase the range of EVs, there are two different ways. The first one is to simply increase the size of the High Voltage Battery (HVB). Unfortunately, this means to increase the size of the most expensive component of an EV, and after all, it is not a very sustainable way. The second way, which is our solution of choice, is to make EVs more efficient.

Kirchhoff's current law states that the sum of all currents at a node of an electric system is equal to 0 A. However, considering measurement signals of nodes in the power trains of EVs with distributed sensor systems, the sum of all currents can differ up to 20 % of the maximum current (see Figure 1). If we look closer at the Root Mean Square Error (RMSE) of the sum of currents $\text{RMSE}(i_{\text{sum}}) = 0.67\%$, we realize that it has the same value as the mean current of the DCDC converter $\mu_{i_{\text{DCDC}}} = 0.67\%$.

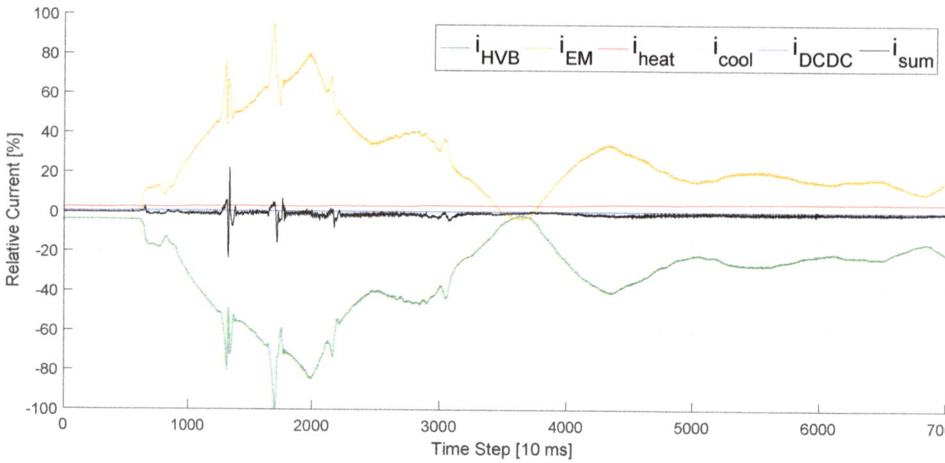

Figure 1. Currents of all HV components in an EV on a test drive. The sum of all currents i_{sum} is plotted in black. According to *Kirchhoff's current law*, it should be constantly 0 %. However, looking at the measurements shows that the deviation i_{sum} is higher than the current of the DCDC converter i_{DCDC}. Even its noise spectrum is approximately half as high as the consumption of the heating i_{heat}, which is the second largest consumer in this drive.

A different value than 0 A for the sum of all currents indicates that there is a divergence between measurements and real values. The divergence becomes problematic when the power train is operating close to the system boundaries. For example, there are boundaries for the protection of the HVB. The HVB is only capable of discharging or charging a restricted amount of power. Higher amounts would threaten the HVB's lifetime and safety [2]. To ensure a safe operation mode even for high divergences between measurements and real values, additional protection offsets (see Figure 2) might be added to the boundaries, although they have some drawbacks.

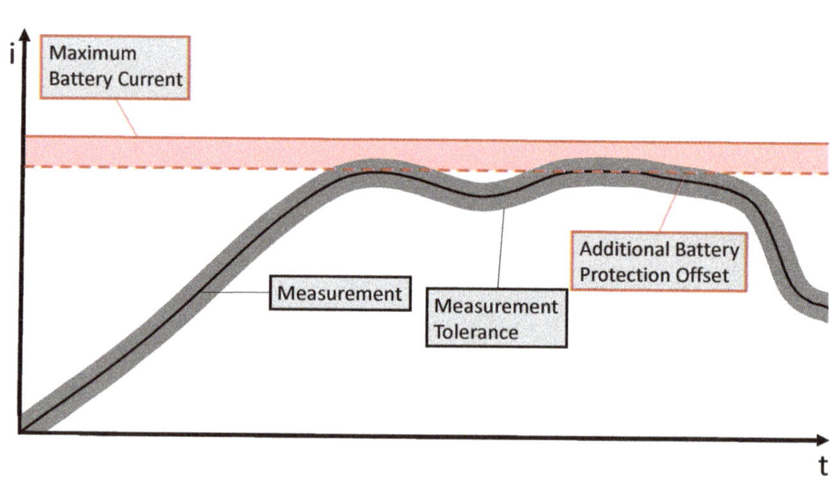

Figure 2. A simplified example of offsets for protection of the HVB. The measured value (black) differs from the real value in the range of some tolerance (grey). To prevent exceeding the battery limit (red, solid) even under the worst measurement conditions, an additional battery protection offset (red, dashed) is introduced. The same principle is used analogously with negative currents. It can be extended to other High Voltage (HV) components.

For example, in the charging case, most notably during recuperation, the HVB might not allow the full power level, even though it would be capable of handling it. Thus, the amount of power charged to the battery is restricted and the EV loses cruising range while its power consumption increases. In the opposite case, the system might not release requested power, although the HVB could provide it in reality. This additional restriction of power decreases the EV's performance. As can be seen from the two examples above, minimizing the magnitude of the protection offsets also allows increasing the performance as the efficiency and the cruising range of EVs.

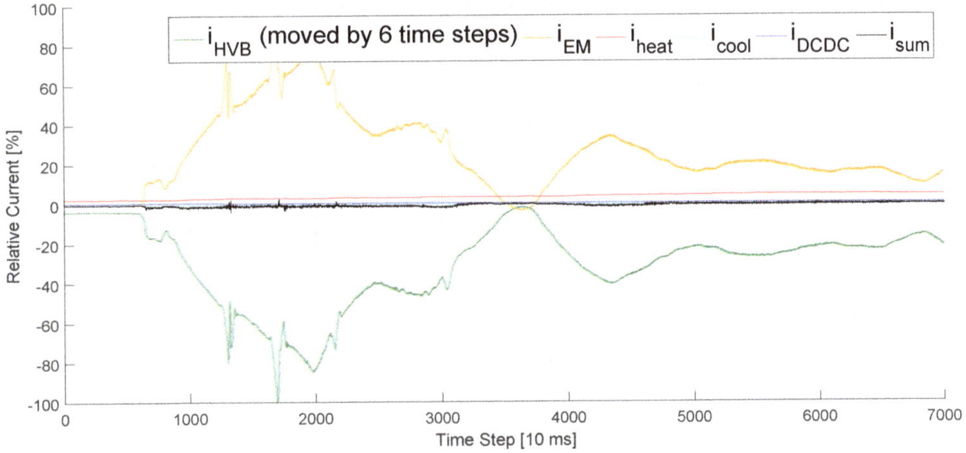

Figure 3. The same test drive as in Figure 1 but with the battery current i_{HVB} (green) shifted by six time steps. The sum of all currents i_{sum} (black) is significantly closer to 0 %.

Besides measurement faults and sensor uncertainties [3], the divergence between measurements and real values is caused by time delays. Figure 3 shows an example of the sum of all currents i_{sum} being reduced by shifting a signal by 6 time steps. The delays result from distributed sensor systems in the power train as plotted in Figure 4. The High Voltage (HV) components have their own Electronic Control Unit (ECU) which is connected with the current sensors and processes the sensor information. The ECUs exchange this information via bus systems. The buses require individual amounts of time to send the measurement signals. Thus, from an ECU's point of view, the sensor information from other ECUs arrives with individual delays (see the Ego ECU in Figure 4). These individual delays could be compensated easily with a synchronized clock and time stamps as part of each bus message. However, this solution would have two drawbacks. First, it would increase the bus traffic as not only the measurement information must be carried by the messages but also the time stamp. As a result, the EV would either require a faster bus which is able to transport more information, or it would have to reduce the information exchanged between the ECUs. Second, there exists no clock in the power trains of modern series EVs which is synchronized with all ECUs at the same frequency as the message exchange. Usually, the ECUs are synchronized in a longer time frame than they communicate. Thus, the time stamp solution would require additional or higher performing hardware and increase the costs for the production of the EV.

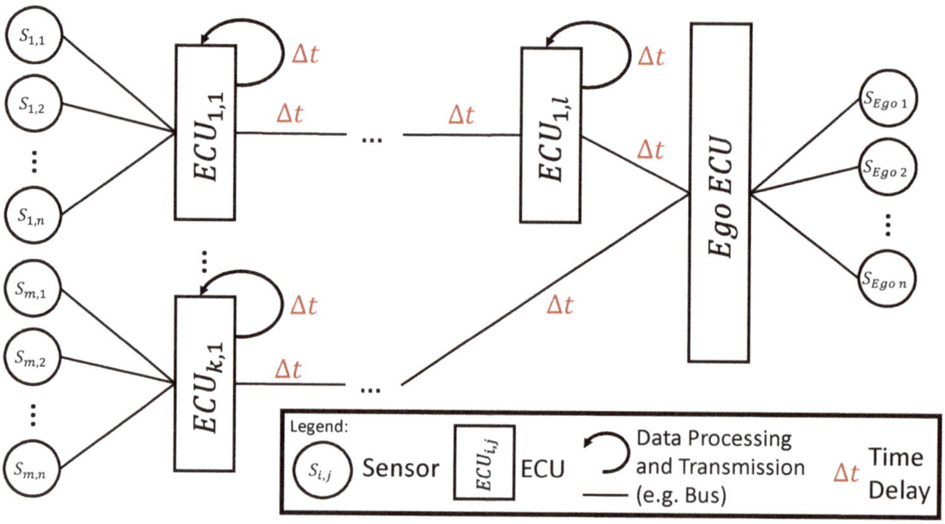

Figure 4. A schematic example of an automotive bus system with higlighted sources of time delays. Please note that the time delays are highly individual and not necessarily equal, but constant or only slowly changing. The ECUs can be connected directly or indirectly via other ECUs. The Ego ECU is not able to reconstruct the time delays, because it only knows the received measurement values and their last sender. It has no further information about the time passed since the measurement's original creation.

The aim of this work is to automatically detect the time delay between measurement signals from different sensors without additional hardware. For this purpose, we develop two different approaches. One of them is based on Linear Regression (LR), whereas the other one optimizes the estimated variance of the difference between several signals. We compare our approaches to other state-of-the-art Time Delay Estimation (TDE) algorithms and evaluate them with a focus on precision and run-time efficiency. Apart from allowing a more accurate power distribution, the automated TDE helps to reduce the battery protection offset and thus to increase the performance, efficiency and cruising range of EVs.

The rest of this paper is structured as follows. Section 2 states related work and the similarities and differences to our work. Furthermore, Section 2 highlights the contributions of our work to the state of the art. In Section 3, we explain the theory behind our work before we describe the practical experiments in Section 4. The experiments' results, stated in Section 5, show us the performance of the algorithms for our use case. Based on this evaluation, we take the best performing algorithm and optimize it further. The optimization steps can be taken from Section 3.4 and their impacts to the results from Section 5.4. In Section 6, we discuss the advantages and drawbacks of all proposed concepts. Finally, we draw our conclusions and give a short outlook in Section 7.

2. State of the Art

There exists plenty of literature about TDE, although—to the best of our knowledge—none of them is tailored to the specific problem of TDE of current signals in EVs. In the following, we present several publications about TDE from different fields of application, such as embedded systems, acoustics, medicine, positioning, aeronautics, process technology, and robotics.

An approach which also deals with EVs and time delays is the one by Guo et al. [4]. However, their approach is similar to ours only at the first look. Their goal is to stabilize a grid of electric sources and sinks with EVs. For the stabilization of the grid, they propose time delay resistent control strategies of smart grids with EVs. The EVs are able to charge bidirectionally. The bidirectional charging is used to smooth disturbances and respond rapidly to fast occurring changes in the power distribution of the grid. An example for such a rapidly occurring change in the times of renewable energies is the power output of wind turbines when a strong wind occurs. Compared to our approach, Guo's focus is rather on the control strategy than on the TDE. Another difference with our work is that Guo's system is rather macroscopic with lots of different elements and many EVs in the grid. Our system is instead quite microscopic. We consider a single EV with a power train of around five sources and sinks. Our communication network might be smaller than the number of HV components as some consumers might share the same ECU. For example, the heating and the cooling component of an EV use both the climate control ECU for bus communication.

Kali et al. [5] propose a controller with TDE for Electric Machines (EMs). The TDE is executed state-based with the help of a model of the EM. The model design demands expert knowledge about the physical principles of an EM. This is justified for Kali et al. as they require the same knowledge for their controller. However, for our case, we want to be able to estimate the time delays without further knowledge about the HV components. Our TDE shall be executable with nothing else than the available measurement data.

Zeng et al. [6] introduce a statistical approach to predict the delay of a bus message. The content of the messages does not need to be known to achieve high accuracy. This is different from our scenario where we want to make use of the information carried by the message. In contrast to Zeng et al., we do not require predicting the time delay accurately to milliseconds. For our purposes, an estimation of the number of delayed discrete time steps is sufficient.

Not from the field of electric power trains or bus communication, but from acoustics is the approach shown by Lourtie and Moura [7]. They use a stochastic approach to model time delays in an acoustic path environment. Like ours, their environment consists of several sources. However, in contrast to our scenario, the delay they want to estimate varies with time. In our case, we assume the time delay to be constant in a short time frame. For longer periods, it might change slowly. The reason for the slowly changing time delay is that it is caused during the wake up procedure of the EV. The ECUs wake up in an unsynchronized way. Afterwards, the ECUs are synchronized on a relatively large time frame (e.g., 1 s), but work based on short time steps (e.g., 10 ms).

Another acoustics application for TDE is shown by He et alii [8]. They use the so-called Multichannel Cross-Correlation Coefficient algorithm to estimate time delays of speech sources in noisy and reverberant environments.

Svilainis et al. [9] present another interesting approach. Their goal is to estimate the time passed between emitting an ultrasonic signal and absorbing its reflection. Like Zeng et al., they require high precision. Another difference to our approach is that their algorithms make use of the pulse form of ultrasonic signals. Our signal as plotted in Figure 1 can vary in a large range and does not necessarily contain pulses (e.g., after time step 5,000).

Mirzaei et al. expand TDE for ultrasonic signals to the field of medicine [10]. The authors introduce a window-based TDE approach to estimate the time passed between two frames of radio-frequency data. They compare the results of the new window-based approach to their previously developed, optimization-based method [11] and to Normalized Cross-Correlation.

Recently, Garcez et al. published their work on a similar problem to ours, but in a completely different field of application [12]. Like bus systems of EVs, Global Navigation Satellite Systemss (GNSSs) systems have real-time requirements. Their goal is to minimize deviations between measurements and real position data. The time delays are caused during the transmission of GNSS messages, when the signals do not take straight lines of sight, but are reflected on their way or suffer from noise. The authors propose a tensor-based subspace tracking algorithm to efficiently estimate time delays of received GNSS signals.

A similar approach is presented by Xie et al. for an indoor positioning sensing system [13]. They sense positions of mobile devices based on the signal strength and the signal's time delay since its transmission from a base station. For the TDE, Xie et al. combine Cross-Correlation with Quadratic Fitting. This is similar to our LR approach (see Section 3.2), where we try to fit the signals with quadratic functions to retrieve the delay between them. Like Garcez et al., they have to deal with the problem that the signals are often reflected and do not take direct lines of sight. Different to Garcez et al., Xie's approach takes the strength of the signal into account for retrieving a more exact position estimation. For our work, we cannot take advantage of this information, because in wire-based bus systems all signals are equally strong.

Schmidhammer et al. estimate positions of moving, non-cooperative objects in vehicular environments [14]. Their idea is to estimate the position of an object based on time delays in a network of distributed receiving and transmitting nodes. In contrast to our work, the networking nodes of Schmidhammer et al. are not necessarily on-board the vehicle, but can also be mounted on the road infrastructure.

Emadzadeh et al. [15] show an inspiring approach for detecting the relative position of spacecrafts. For retrieving the position, they examine an X-ray signal received by two spacecrafts and determine the time delay between them. For the TDE, they use Adaptive Filters (AFs). This approach seems very promising to us. We implement the algorithms of Emadzadeh et al. and compare them to ours in order to find out if their approach can be transferred from X-ray signals to current measurements in the power train of EVs.

Like Emadzadeh et al., Liu et al. focus on AFs [16]. Compared to our problem of fixed or only slowly changing time delays, the difference in Liu et al. is that they deal with time-varying time delays. That makes further processing steps necessary. For example, they require a transition probability matrix and an initial probability distribution vector to model the time delay changes with a Markov chain.

Park et al. analyze time series data with Autoencoders and Long Short-Term Memory Neural Networks (LSTMs) to detect faults in industrial processes [17]. The authors emphasize the importance of TDE for correct fault detection. However, they focus only on time delays caused by their own fault detection system. Our focus lies on earlier steps in the processing chain. We want to detect time delays between the input signals before they are passed to other computation processes. Furthermore, we want to implement algorithms which are able to learn on-board the automotive ECUs and adapt themselves to new data. As the training of Neural Networks is quite memory intensive and demands high computational power, they do not belong to our methods of choice.

Close to the application field of industrial processes is the approach of Srinivasa Rao et al. [18]. In their recent article, the authors propose fuzzy parametric uncertainty to mathematically model systems with time delays. Their goal is to enable a robust controller design. For this purpose, they first approximate the time delay system as an interval system. After retrieving the intervals, they design an optimal controller for these. Like Guo et al., Srinivasa Rao et al. focus on how to retrieve an optimal controller, which is not part of our work. Although they focus on the control of industrial processes, their article is very general. Besides industrial plants, they also mention potential fields of application, such as EMs or robot manipulators.

Time delay compensation for robots is the focus of Shen et al [19]. Their focus is on teleoperating robots which require knowledge about the time delay between the master and the slave robot for stable operation. The robots and their communication channels are modeled as extended dynamical system. For this system, Shen et al. develop a cascade observer which is able to control it in a stable way. The authors assume that a sufficiently accurate value for the TDE is given and concentrate on its compensation. This is different to our work here. We explicitly want to estimate the time delay.

You et al. develop a proportional multiple integral observer for fuzzy systems [20]. The goal of their work is the same as ours. They want to minimize deviations between measurements and real values caused by time delays and measurement inaccuracies. Their time delays are also varying. Unlike the varying time delays presented before, the ones of You et al. do not vary with time but rather with states. Their focus is also on industrial processes and not on electric power trains. However, the main difference between our works is that You et al. want to minimize time delays and measurement inaccuracies with the same system.

Our approach follows the *divide and conquer* strategy and faces the two problems separately. We focus on the problem of measurement deviations caused by measurement inaccuracies in our previous work [3]. However, measurement inaccuracies are not part of this work. Here, we assume that the measurements are appropriately accurate and that the main deviations are caused by time delays as shown in Figure 1 and Figure 3. Thus, TDE is our solution of choice to minimize the deviations.

Our contribution in this article is the development of a regression-based approach and an algorithm based on Variance Minimization (VM) for TDE as first presented in [21]. We transfer the ideas introduced by Emadzadeh et al. to the domain of currents in the HV system of EVs and compare the results to our approaches in matters of accuracy and computational performance. Our TDE works only with the data available in modern series EVs and does not require an additional clock. In addition to [21], we introduce an optimization of the most accurate and efficient of our evaluated approaches. We further evaluate the optimization both on artificially created data with known ground truth as well as real drive data with unknown ground truth.

3. Concepts

In this section, we introduce the algorithms and shortly explain the concepts from other authors which we implement and compare for TDE. From now on, for the sake of easier understanding, we focus on the current of the EM i_{EM} and the HVB i_{HVB} (without other consumers than the EM) as examples. Nevertheless, the proposed methods can be extended to every current signal in the HV system of an EV. Furthermore, we inverse the sign of i_{HVB} from now on to make its shape similar to the one of the EM. Thus, we can treat the HVB current signal as a delayed or preceded version of the EM, respectively.

Our goal is to find the time delay t_d in a bus system which can be described as

$$\begin{aligned} x_1(t) &= i_1(t) + n_1(t) \\ x_2(t) &= i_2(t - t_d) + n_2(t - t_d), \end{aligned} \tag{1}$$

where t stands for the time step, $x_1(t)$ is the measurement signal of the faster component, $x_2(t)$ describes the slower component's signal, $i_1(t)$ and $i_2(t)$ describe the corresponding currents and $n_1(t)$ and $n_2(t)$ are noise terms [15]. As we cannot retrieve the currents $i_1(t)$ and $i_2(t)$ directly, we cannot minimize the difference between $i_1(t)$ and $i_2(t)$. Instead, we directly minimize the difference between the two measurement signals $x_1(t)$ and $x_2(t)$.

3.1. Adaptive Filter

The idea of Emadzadeh et al. is to model the time delay as Finite Impulse Response (FIR) filter. They define $x_1(t)$ to be the faster signal. For each measurement $x_2(t_i)$ at time step t_i, they collect a row of the last M measurements of the other signal

$$\mathbf{x}_1(t_i - M + 1 : t_i) = [x_1(t_i - M + 1), x_1(t_i - M + 2), \dots x_1(t_i - 1), x_1(t_i)]\,. \tag{2}$$

Then, the authors search for an optimal channel impulse response vector ω^* such that the deviation between $x_2(t_i)$ and $\mathbf{x}_1(t_i - M + 1 : t_i)\omega^*$ becomes minimal. Mathematically, this can be expressed by the minimization of the expectation value of the Mean Squared Error (MSE) between the measurement value of the slower signal and the filtered measurement row of the faster signal. It results in the formula

$$\omega^* = \underset{\omega}{\operatorname{argmin}}\, E\left[(x_2(t_i) - \mathbf{x}_1(t_i - M + 1 : t_i)\omega)^2\right]. \tag{3}$$

This is similar to our VM approach (see Section 3.3) with the difference that we minimize the variance instead of the MSE. In Emadzadeh's work, the optimal factor ω^* is estimated recursively. For the recursion, the authors implement and compare the four algorithms Least Mean-Squares (LMS), Normalized Least Mean-Squares (NLMS), Least Mean-Fourth (LMF) and Recursive Least-Squares (RLS). The optimal time delay estimate t_d^* is then the one where the impulse response $\boldsymbol{\omega}^*$ reaches its maximum, or mathematically

$$t_d^* = \underset{i\in[1,M]}{\operatorname{argmax}}\ \boldsymbol{\omega}^*(i) - 1. \tag{4}$$

For further details on this approach, we kindly refer the interested reader to [15].

3.2. Linear Regression

Our first approach is to use LR to identify the *basis functions* of two received signals and compare the horizontal offset between these functions. As degree of the basis function, we choose a parabola for two reasons. First, the sampling frequency of our measurements is high enough to fit the signals with a parabola for a short time duration. Second, the comparison of the horizontal offset is easiest with a parabola because it only has one extremum.

We collect the last M measurements x_k of the HV components $k \in \{\mathrm{EM}, \mathrm{HVB}\}$ in a measurement vector $\boldsymbol{y_k} = \begin{pmatrix} x_k(t) & x_k(t-1) & \cdots & x_k(t-M+1)\end{pmatrix}$. Then, we retrieve the weight vector $\boldsymbol{w_k} = \begin{pmatrix} w_{k,0} & w_{k,1} & w_{k,2}\end{pmatrix}^T$ with LR [22] according to

$$\boldsymbol{w_k} = \left(\sum_{n=1}^{N} \boldsymbol{\phi}^n (\boldsymbol{\phi}^n)^T\right)^{-1} \sum_{n=1}^{N} y_k^n \boldsymbol{\phi}^n. \tag{5}$$

Here, the notation y_k^n and $\boldsymbol{\phi}^n$ represents the n-th column of $\boldsymbol{y}_k$ and $\boldsymbol{\phi}$, respectively. The so-called *design matrix*

$$\boldsymbol{\phi} = \begin{pmatrix} 1 & t & t^2 \\ 1 & t-1 & (t-1)^2 \\ \vdots & \vdots & \vdots \\ 1 & t-M+1 & (t-M+1)^2 \end{pmatrix}^T$$

consists of $N = 3$ columns in our case.

With the weight vector from (5), we are able to fit a parabola

$$f_k(t) = w_{k,0} + w_{k,1}t + w_{k,2}t^2 \tag{6}$$

as basis function to the measurement vector $\boldsymbol{y_k}$.

After retrieving the basis functions in (6), we transfer them into vertex form

$$f_k(t) = w_{k,2}(t - x_{k,\text{vertex}})^2 + y_{k,\text{vertex}} \tag{7}$$

to identify the coordinates $(x_{k,\text{vertex}}, y_{k,\text{vertex}})$ of each basis function's vertex. The estimated time delay between the EM and the HVB current signals is then given by the difference on the time axis between their vertices according to

$$t_d^* = x_{\text{EM, vertex}} - x_{\text{HVB,vertex}}. \tag{8}$$

3.3. Variance Minimization

Our second approach is to minimize the variance of the difference between two signals $x_1(t)$ and $x_2(t)$ by shifting the signal $x_2(t)$ forward.

Like in Section 3.2, we collect the two signals $x_1(t)$ and $x_2(t)$ for M time steps. A straightforward idea for the minimization of the difference between $x_1(t)$ and $x_2(t)$ is to minimize their estimated MSE

$$\mathrm{MSE}(x_1(t), x_2(t)) = \frac{1}{M - T_{\max}} \sum_{t=1}^{M-T_{\max}} (x_1(t) - x_2(t + t_{d,i}))^2 \tag{9}$$

with different time shifts $t_{d,i}$ in a pre-defined range $t_{d,i} \in [T_{\min}, T_{\max}]$ with $T_{\max} < M$. However, our experiments show that we need a relatively high M to achieve stable results. We can significantly minimize M, if we take the estimated expected value

$$E = \frac{1}{M - T_{\max}} \sum_{t=1}^{M-T_{\max}} (x_1(t) - x_2(t + t_{d,i})) \,. \tag{10}$$

into account. Thus, instead of minimizing the MSE from (9), we minimize the estimated variance of the difference between the two signals

$$\sigma^2 = \frac{1}{M - T_{\max}} \sum_{t=1}^{M-T_{\max}} ((x_1(t) - x_2(t + t_{d,i})) - E)^2. \tag{11}$$

The time delay between the EM and the HVB current signals is the $t_{d,i}$ that minimizes the variance

$$t_d^* = \underset{t_{d,i}}{\mathrm{argmin}}\, \sigma^2. \tag{12}$$

As we do not know in the beginning whether $x_{\mathrm{EM}}(t)$ or $x_{\mathrm{HVB}}(t)$ is the faster signal, we have to choose one of them as $x_1(t)$ and the other one as $x_2(t)$ for the first execution and try $T_{\min} = -T_{\max}$. From the second execution on, the value of $T_{\min}$ and $T_{\max}$ can be reduced and chosen recursively, because the EV's bus system usually changes its time delay only once in the beginning, but not during advanced execution. Therefore, we choose $T_{\min}(t) = t_d(t-1) - 1$ and $T_{\max}(t) = t_d(t-1) + 1$ from the algorithm's second execution on.

3.4. Optimized Variance Minimization

As the results of our experiments (see Section 5) show, the VM concept provides the best results in terms of RMSE, run-time and required frame size. However, when running this concept in real time (both on simulated and real data, see Section 5.4), we find that the TDE is unstable and that the estimated time delay frequently alternates between different values. These many changes of the estimated time delay contradict the fact that the time delay is rather stable in reality, and that, if any, changes occur after relatively long periods. Thus, in order to stabilize the TDE, we suggest the following improvement of the *plain* VM approach from Section 3.3.

The main idea of the stabilization is to use a statistical test. The test's purpose is to quantify the *reliability* of the input data segment on which the TDE is performed. In fact, we know from Section 3.3 that the estimated time delay t_d at time step t minimizes the variance given by Equation (11). This equation in turn is based on the M last values of both signals. Due to the noise in the data, the estimation for the next step can jump to a different value, even if the vast majority of data points $(M - 1)$ are shared between the two steps. The idea is thus to compare at each step the minimal variance with the second smallest one. If the difference between both in relative terms is not sufficiently large, we presume that the TDE is not reliable, and consequentially do not estimate a time delay. In this case, we simply keep the prediction from the last step. Otherwise, we update the estimation to the newly calculated t_d.

Formally, at each time step, we calculate the variance criterion from Equation (11) for each potential time delay $t_{d,i}$. Let us denote this by $\sigma^2(t_{d,i})$. Then, we know that the least achievable variance is given by

$$\sigma^2(t_d^*) = \min_{t_{d,i}} \sigma^2(t_{d,i}). \tag{13}$$

The second smallest achievable variance in turn is given by

$$\sigma^2(t_d^{**}) = \min_{t_{d,i} \neq t_d^*} \sigma^2(t_{d,i}). \tag{14}$$

In other words, we minimize the variance over all potential time delays except that which minimizes it (t_d^*). By definition, we have $\sigma^2(t_d^{**}) \geq \sigma^2(t_d^*)$. The intuition is that if the difference between those two values is not large enough, the noise makes it impossible to tell with high confidence which one is the real minimizer. The minimum in Equation (13) might result in t_d^* by random noise instead of being the true minimum. Thus, we suggest deciding whether to perform an update based on the criterion

$$\frac{\sigma^2(t_d^{**}) - \sigma^2(t_d^*)}{\sigma^2(t_d^*)} > K, \tag{15}$$

where K is a hyper-parameter defining the minimal percentage error required to perform an update. Clearly, the larger K, the more severe is the criterion, and the fewer updates are done. Therefore, we choose K to strike a balance between reliability on the one hand, and being up-to-date on the other hand. In fact, if we choose K too high, updates are performed only rarely, so that we can miss changes in the underlying real time delay. If K is chosen too small, then the predictions are more unstable. We empirically found $K = 0.2$ to strike a balance between both criteria for our power train data.

4. Experimental Setup

In this section, we explain the data and the setup for the experiments to evaluate the performance of the three concepts for TDE and the optimization presented above.

4.1. Data

For the evaluation of the three concepts and the optimization shown in Section 3, we use 74 data sets. The data sets contain all currents of the HV system and are recorded during representative drives on public roads with close to production EVs. We use bus loggers to record the data. The loggers store the received measurement signals from all ECUs and write them to a log file during each time step. After driving, we use the log files to execute our experiments and evaluate our approaches. Thus, the algorithms get at each time step the same input data which they would receive during execution on an ECU in the real EV. The recordings correspond to 10 h 33 min of driving. For the experiments, the 74 data sets are divided into 409 sub-data sets with a maximum length of 10,000 time steps. The minimum length among the 409 sub-data sets is 1,807 time steps.

For the Optimized VM approach, we create an additional data set artificially. The artificial data set bases upon the real data sets described above. However, instead of computing the time delay between two real signals, we introduce an artificial signal. This artificial signal is a real signal shifted by some time steps. We can then compute the time delay between the original signal and its artificially delayed correspondence. This has the advantage that we exactly know the time delay and thus know the ground truth.

4.2. Experiments

According to *Kirchhoff's current law*, we assume that the sum of the measurements of i_{HVB}, i_{EM}, i_{heat}, i_{cool} and i_{DCDC} is zero. Thus, we estimate the *ground truth* of the time delay for each real data set by minimizing the MSE of the complete data set (see Equation (9)). In this case, M is the length of the data set, T_{max} is 10 time steps and T_{min} is -10 time steps, since the time delay in the EV is normally smaller than ten time steps.

In the experiment of the AF concept, we choose the frame size of 2^{10} as proposed by [15]. Additionally, we execute our experiments with a more efficient frame size of 2^8. We assume the length of the AF to be 10. All the other parameters for the used algorithms are chosen equally to [15]. Furthermore, we evaluate five different learning rates for LMS.

For the evaluation of the VM concept we test different frame sizes $M \in \{30, 50, 100, 200, 300\}$. In the first calculation, we also choose $T_{max} = 10$ and $T_{min} = -10$. From the second execution on we select $T_{min}(t) = t_d(t-1) - 1$ and $T_{max}(t) = t_d(t-1) + 1$.

For the Optimized VM, we choose a fixed frame size of 50 time steps. This frame size proved to be the best compromise between run-time and accuracy in previous experiments as described in subsection 5.3.

For each of the three concepts, we calculate the time delay every 20th time step. In total, this results in around 90,000 time delay estimations for each concept.

4.3. Environment

All concepts are implemented in Matlab R2015b with Microsoft Windows 10 on an HP® EliteBook™840 G3 with an Intel® Core™i5-6300U 2.40GHz CPU and 8 GB RAM.

5. Results

In this section, we present the results of our experiments and evaluate the performance of the three concepts and the optimization individually. The results of all three algorithms compared next to each other can be found in the next section.

5.1. Adaptive Filter

Based on the learning rate and parameters in [15], the RLS algorithm performs better than the LMS, NLMS and LMF algorithms (see Table 1).

Table 1. RMSE and run-time analysis of the AF concept for different algorithms with a frame size of 2^{10}.

	RMSE	Average Run-Time (s)
LMS	2.8972	$1.13 \cdot 10^{-2}$
NLMS	2.5163	$1.31 \cdot 10^{-2}$
LMF	2.6138	$1.14 \cdot 10^{-2}$
RLS	2.276	$1.07 \cdot 10^{-2}$

Furthermore, we analyze the learning rate for the LMS algorithm. As mentioned in [15], the learning rate μ is typically chosen in the range $0 < \mu < 2/(M\sigma_u^2)$, where σ_u^2 is the input signal variance and M is the length of the filter. Thus, we compare the performance of LMS with different $\mu = a/(M\sigma_u^2)$ and $a \in \{0.01, 0.05, 0.1, 0.5, 1\}$. In Table 2, the RMSE has the minimal value of 2.1479 with $a = 0.1$. It is much smaller than the RMSE of 2.8972 with the fixed learning rate in [15]. For a too large or a too small a, the performance of the LMS decreases significantly. This result is expected, since a too small learning rate leads to slow convergence while a large one most often misses the optimum.

Table 2. RMSE and run-time analysis of LMS with different learning rates and a frame size of 2^{10}.

	RMSE	Average Run-Time (s)
$a = 0.01$	3.0451	$1.14 \cdot 10^{-2}$
$a = 0.05$	2.3608	$1.14 \cdot 10^{-2}$
$a = 0.1$	2.1479	$1.14 \cdot 10^{-2}$
$a = 0.5$	2.6484	$1.14 \cdot 10^{-2}$
$a = 1$	3.2474	$1.14 \cdot 10^{-2}$

In addition, we evaluate the LMS algorithm with a more efficient frame size of 2^8. As printed in Table 3, the smaller frame size improves the run-time. Although some non-optimal learning rates improve their estimation accuracy, which we explain with the drop of local minima due to the shortened frame, the two best learning rates in the experiment with the frame size of 2^{10} increase their estimation errors with the smaller frame size.

Table 3. RMSE and run-time analysis of the LMS algorithm with a shorter frame size of 2^8.

	RMSE	Average Run-Time (s)
$a = 0.01$	2.8603	$2.80 \cdot 10^{-3}$
$a = 0.05$	2.5598	$2.80 \cdot 10^{-3}$
$a = 0.1$	2.3994	$2.80 \cdot 10^{-3}$
$a = 0.5$	2.4983	$2.80 \cdot 10^{-3}$
$a = 1$	3.0781	$2.80 \cdot 10^{-3}$

5.2. Linear Regression

Our first approach LR is, to a large extent, affected by noise and the offset between the two signals caused by measurement inaccuracies. Especially this offset leads to an imprecise estimation of the vertices and thus a wrong estimated time delay. Figure 5 shows an example for such a wrong estimation. In this data set, the time delay between i_{HVB} and i_{EM} is equal to 6 time steps. We train both curves on 200 measurement samples of their corresponding signal. However, due to noise and some vertical offset between the signals the vertex of the slower signal is not only shifted to the right but also to the top. The shift in vertical direction also affects the horizontal position of the vertex and results in a wrong TDE of 43 time steps.

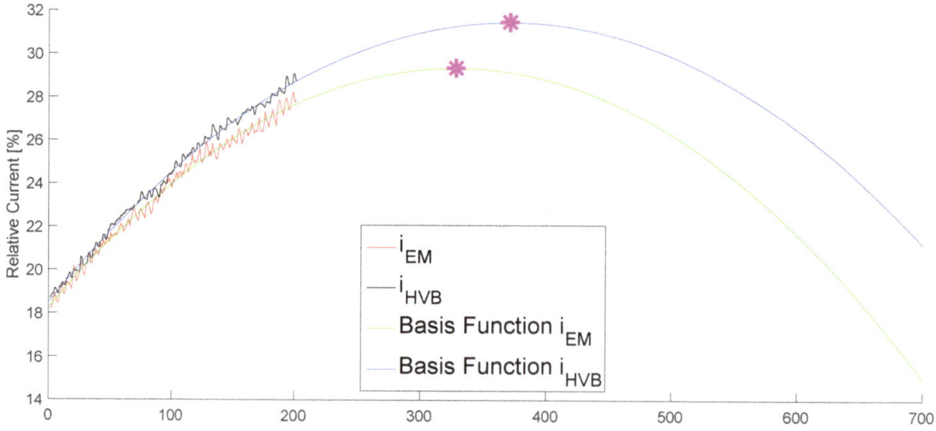

Figure 5. Basis functions of i_{HVB} (black) and i_{EM} (red) simulated by LR (blue and green, respectively). The magenta marked points are the vertexes. Their horizontal difference is 43 time steps in contrast to the real time delay which is 6 time steps. The wrong TDE is caused by the noise and the vertical offset of the measurement signals.

Table 4 shows the results and the average run-time of this approach with three different frame sizes. The run-time grows with increasing frame sizes, whereas the RMSE becomes smaller. Nevertheless, the RMSEs are in general very high even for large frames.

Table 4. RMSE and run-time analysis of the LR concept.

	RMSE	Average Run-Time (s)
Frame Size 30	$5.83 \cdot 10^{10}$	$1.74 \cdot 10^{-4}$
Frame Size 200	$4.92 \cdot 10^{4}$	$9.80 \cdot 10^{-4}$
Frame Size 300	$4.47 \cdot 10^{4}$	$1.40 \cdot 10^{-3}$

5.3. Variance Minimization

Table 5 shows the RMSE between the ground truth of the time delay and the calculated time delay. Furthermore, the table shows the average of the run-time for each time delay calculation, corresponding to different frame sizes M (in Equation (11)). We see that the concept requires a relatively short run-time as it benefits from the recursive calculation only in the area $t_{d,i} \in [T_{\min}, T_{\max}]$ with $T_{\min}(t) = t_d(t-1) - 1$ and $T_{\max}(t) = t_d(t-1) + 1$. In addition, the RMSE decreases while the size of the frame increases. The accuracy has a large improvement when the frame is enlarged from 30 time steps to 50 time steps.

Table 5. RMSE and run-time analysis of the VM concept.

	RMSE	Average Run-Time (s)
Frame Size 30	2.0696	$4.54 \cdot 10^{-5}$
Frame Size 50	1.3034	$4.70 \cdot 10^{-5}$
Frame Size 100	1.2364	$4.77 \cdot 10^{-5}$
Frame Size 200	1.2215	$5.16 \cdot 10^{-5}$
Frame Size 300	1.1825	$5.79 \cdot 10^{-5}$

5.4. Optimized Variance Minimization

We evaluate the proposed stabilization approach twofold. First, we evaluate it based on simulated data with known ground truth time delay. Second, we evaluate the approach on real signals. While the first experiment shows the accuracy of the proposed approach, the second one shows its effectiveness in providing more stability.

5.4.1. Evaluation with Simulated Signals

In this experiment, we first take a current signal $x_1(t)$ from a real-world data set recorded on-board of an EV. Based on x_1, we then create a second signal $x_2(t) = x_1(t - t_d(t)) + n_2(t)$. Therefore, the ground-truth time delay $t_d(t) \geq 0$ is a realization of a random jump process that in known in advance. Furthermore, $n_2(t)$ is a white noise process whose variance is chosen such that the resulting Signal-to-Noise Ratio (SNR) is equal to -10. We then run our VM algorithm with and without stabilization to detect the delay $t_d(t)$. Figure 6 shows the results of this experiment.

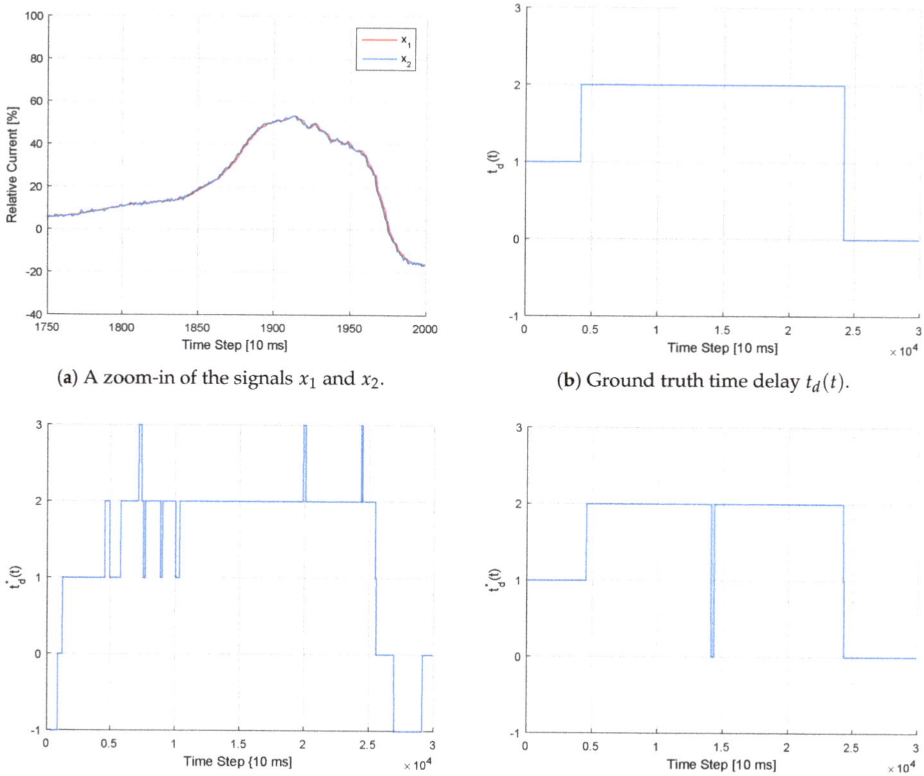

(**a**) A zoom-in of the signals x_1 and x_2.

(**b**) Ground truth time delay $t_d(t)$.

(**c**) Estimated time delay without stabilization.

(**d**) Estimated time delay with stabilization.

Figure 6. Illustration of the TDE estimation procedure using the VM approach with and without stabilization for the case of simulated signals. Clearly, the estimation is more stable when using the criterion in equation (15). Waiting for the *right* moment to perform an update comes however with the expense of a slightly delayed, yet more reliable, prediction. For example, the jump of $t_d(t)$ from 1 to 2 was detected with a delay of around 320 steps, which corresponds to around 3.2 seconds.

5.4.2. Evaluation with Real Signals

In this experiment, we take both signals $x_1(t)$ and $x_2(t)$ from a real-world data set. We then run the VM approach with and without stabilization, and plot the results in Figure 7.

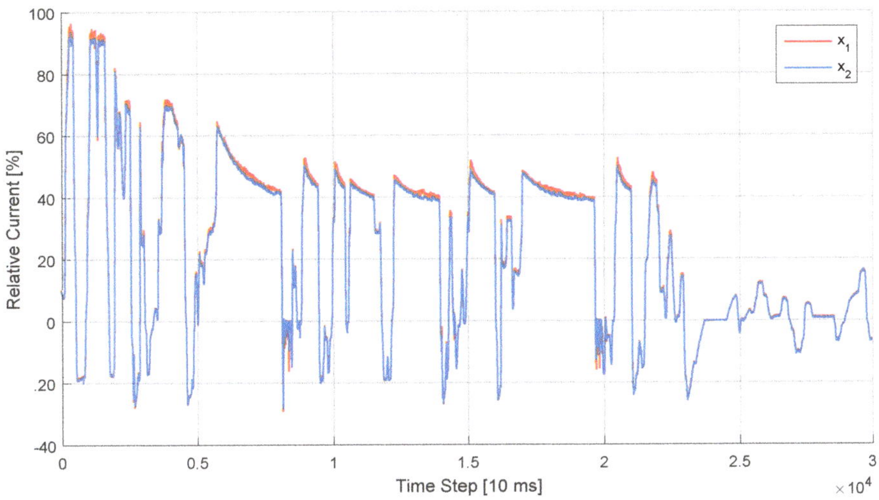

(a) The signals x_1 and x_2.

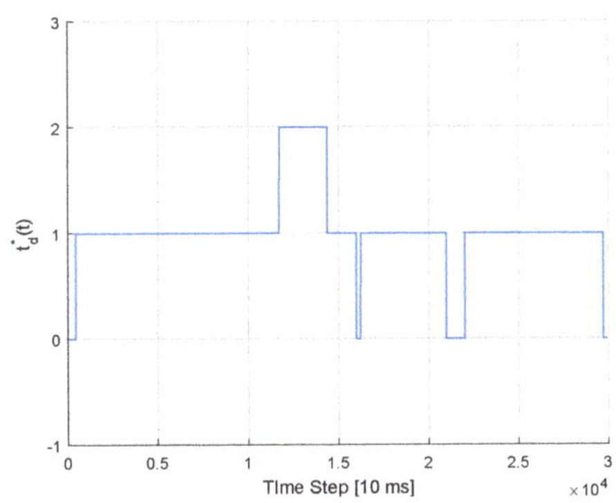

(b) Estimated time delay without stabilization.

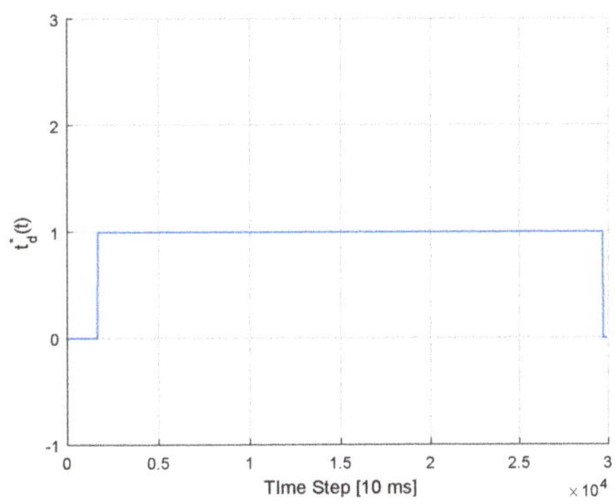

(c) Estimated time delay with stabilization.

Figure 7. Illustration of the TDE estimation procedure using the VM approach with and without stabilization for the case of real signals. Although we do not know the underlying true time delay, it is again clear that the TDE is more stable using the suggested approach.

6. Discussion

We discuss the advantages and drawbacks of the previously described and evaluated concepts in this section.

Although it is not as efficient and accurate as VM, the AF approach still retrieves better results than LR. The best results for AFs, in our case, are reached with the LMS algorithm and a learning rate of $\mu = 0.1/(M\sigma_u^2)$ (see Table 2). The learning rate, which must be chosen manually, is one drawback of this algorithm. It can lead to sub-optimal learning if the user chooses a wrong value. In contrast with LMS, the RLS algorithm does not require a learning rate. However, we see that the RLS algorithm has lower accuracy, requires longer run-time and more memory for a larger frame than the VM concept (see Table 6).

Table 6. RMSE, run-time analysis and frame size of all three concepts compared to each other.

	RMSE	Average Run-Time (s)	Frame Size
Adaptive Filter	2.3994	2.80×10^{-3}	2^8
Linear Regression	4.92×10^4	9.80×10^{-4}	200
Variance Minimization	1.3034	4.70×10^{-5}	50

LR has the advantage that it can directly find out the faster component. Thus, one single execution during the same time step for the same signal is sufficient even in the beginning, which makes it interesting, if a computational effective approach is needed. However, its efficiency suffers from the matrix inversion in Equation (5). Even worse, it is the least accurate of the three proposed concepts due to noise and vertical offsets between the signals. The high estimation errors make this approach unfeasible for our purpose. Another drawback is that a parabola is not always the optimal basis function for the regression of measurement signals.

The VM approach does not require such a basis function. Unfortunately, it is not able to detect the faster signal without trying each possible time delay for both signals. This results in a computationally expensive brute force calculation in the first time step. Afterwards, it is very efficient because it must only execute basic math operations and searches only for a restricted number of possible delays. Compared to the other approaches, VM requires the smallest frame size to retrieve feasible results. In total, Table 6 shows clearly that VM is the most accurate and fastest of the three proposed approaches with the lowest memory consumption.

For the high precision and low run-time, we decide to continue our work with the VM concept. Before we are able to apply our approach to series production EVs, we require further optimization as shown in Section 3.4 to stabilize the estimated time delay. This stabilization comes with another drawback. The algorithm requires more time steps to pass before it adapts to a new delay. Nevertheless, regarding that succeeding power train control functions require stable inputs, this drawback seems acceptable for us. Another drawback of the optimization is the threshold value K in Equation (15) which must be chosen manually. Although it does not require expert knowledge but can be set by trial and error, we would prefer an automated way for finding the optimal value for K.

7. Conclusions and Outlook

This article presents three different approaches for TDE of measurement signals in the power train of EVs. As automotive ECUs are designed very efficiently, our evaluation's focus lies also on computation and memory complexity and not solely on accuracy. Unfortunately, LR is not suited for our purposes because it suffers too much from vertical offsets in the measurement data. However, with VM, we present a feasible approach for TDE of distributed sensor systems of EVs. AFs are also not suited because they require too large frame sizes and have lower accuracy than VM. We recommend using VM due to its high estimation accuracy and computational efficiency. As the output of VM is not stable enough to directly process it to power train control functions of series EVs, we optimize it first. For the optimization, we introduce a threshold value as additional requirement for changing the value of the estimated time delay. The new requirement decelerates the detection of changed time delays. Nevertheless it improves the TDE's stability and accuracy.

After the introduction of an automated TDE system, we now know each signal's delay. However, if we correct the delay, we move some signals to the past and lose the measurements corresponding to the latest time steps. This is correct because in fact we do not receive up-to-date measurements, only delayed ones from the past. We really do miss the last measurements and there is a gap between the last received measurement and the present time step. Thus, our next work focuses on possible ways to close this gap by replacing the hidden information about the missing measurements from the latest time steps.

8. Patents

The TDE for the power trains of EVs is registered at the German Patent and Trade Mark Office (DPMA) as patent application. Both the VM approach as well as its RMSE-based version for TDE in the power trains of EVs are registered there as a common patent application. The optimization is registered as a third patent application resulting from this work.

Author Contributions: conceptualization, J.P.; methodology, J.P., X.W. and A.A.; software, J.P. and X.W.; validation, X.W.; formal analysis, J.P. and X.W. investigation, J.P. and X.W.; resources, J.P. and X.W.; data curation, J.P.; writing–original draft preparation, J.P., X.W. and A.A.; writing–review and editing, J.P. and X.W.; visualization, J.P., X.W. and A.A.; supervision, J.P.; project administration, J.P.; funding acquisition, J.P. All authors have read and agreed to the published version of the manuscript.

Funding: This research received no external funding.

Conflicts of Interest: The authors declare that the publication of data must be approved by the BMW Group. Besides that, there are no conflicts of interest. The funders had no role in the design of the study; in the collection, analyses, or interpretation of data; in the writing of the manuscript, or in the decision to publish the results.

Abbreviations

The following abbreviations are used in this manuscript:

AF	Adaptive Filter
AR	Auto-Regressive
ARIMA	Auto-Regressive Integrated Moving Average
ECU	Electronic Control Unit
EM	Electric Machine
EV	Electric Vehicle
FCHEV	Fuel Cell Hybrid Electric Vehicle
FIR	Finite Impulse Response
GNSS	Global Navigation Satellite Systems
HV	High Voltage
HVB	High Voltage Battery
I	Integrated
IDAACS	IEEE International Conference on Intelligent Data Acquisition and Advanced Computing Systems: Technology and Applications
LSTM	Long Short-Term Memory Neural Network
MA	Moving Average
MOS-ELM	Online Sequential Extreme Learning Machine with Memory principle
LMF	Least Mean-Fourth
LMS	Least Mean-Squares
LR	Linear Regression
MSE	Mean Squared Error
NLMS	Normalized Least Mean-Squares
N4SID	Numerical State Space Subspace System Identification
RLMS	Recursive Least Mean-Squares
RLS	Recursive Least-Squares
RMSE	Root Mean Square Error
SNR	Signal-to-Noise Ratio
PCA	Principle Component Analysis
TDE	Time Delay Estimation
VM	Variance Minimization

References

1. Dixon, J.; Anderson, P.B.; Bell, K.; Træholt, C. On the ease of being green: An investigation of the inconvenience of electric vehicle charging. *Appl. Energy* **2020**, *258*. doi:10.1016/j.apenergy.2019.114090. [CrossRef]
2. Komarnicki, P.; Haubrock, J.; Styczynski, Z.A. Electrical components of an EV (Elektrische Komponenten des E-Kfz). In *Elektromobilität und Sektorenkopplung*; Springer: Berlin/Heidelberg, Germany, 2018; pp. 61–109.
3. Pfeiffer, J.; Wolf, P.; Pereira, R. A Fleet-Based Machine Learning Approach for Automatic Detection of Deviations between Measurements and Reality. In Proceedings of the 2019 IEEE Intelligent Vehicles Symposium (IV), Paris, France, 9–12 June 2019; pp. 2086–2092.
4. Guo, Y.; Zhang, L.; Zhao, J.; Wen, F.; Salam, A.; Mao, J.; Li, L. Networked Control of Electric Vehicles for Power System Frequency Regulation with Random Communication Time Delay. *Energies* **2017**, *10*, 621. doi:10.3390/en10050621. [CrossRef]

5. Kali, Y.; Ayala, M.; Rodas, J.; Saad, M.; Doval-Gandoy, J.; Gregor, R.; Benjelloun, K. Current Control of a Six-Phase Induction Machine Drive Based on Discrete-Time Sliding Mode with Time Delay Estimation. *Energies* **2019**, *12*, 170. doi:10.3390/en12010170. [CrossRef]
6. Zeng, H.; Di Natale, M.; Giusto, P.; Sangiovanni-Vincentelli, A. Statistical analysis of Controller Area Network message response times. In Proceedings of the 2009 IEEE International Symposium on Industrial Embedded Systems, Lausanne, Switzerland, 8–10 July 2009; pp. 1–10.
7. Lourtie, I.M.G.; Moura, J.M.F. Optimal Estimation of Time-Varying Delay. In Proceedings of the 1988 International Conference on Acoustics, Speech, and Signal Processing (ICASSP), New York, NY, USA, 11–14 April 1988.
8. He, H.; Chen, J.; Benesty, J.; Zhang, W.; Yang, T. A class of multichannel sparse linear prediction algorithms for time delay estimation of speech sources. *Signal Process.* **2020**, *169*, 107395. doi:10.1016/j.sigpro.2019.107395. [CrossRef]
9. Svilainis, L.; Aleksandrovas, A.; Lukoseviciute, K.; Eidukynas, V. Investigation of the Time of Flight estimation errors induced by neighboring signals. In Proceedings of the 2013 IEEE 7th International Conference on Intelligent Data Acquisition and Advanced Computing Systems (IDAACS), Berlin, Germany, 12–14 September 2013; pp. 413–418.
10. Mirzaei, M.; Asif, A.; Rivaz, H. Combining Total Variation Regularization with Window-Based Time Delay Estimation in Ultrasound Elastography. *IEEE Trans. Med. Imaging* **2019**, *38*, 2744–2754. doi:10.1109/TMI.2019.2913194. [CrossRef] [PubMed]
11. Hashemi, H.S.; Rivaz, H. Global Time-Delay Estimation in Ultrasound Elastography. *IEEE Trans. Ultrason. Ferroelectr. Freq. Control* **2017**, *64*, 1625–1636. doi:10.1109/TUFFC.2017.2717933. [CrossRef] [PubMed]
12. Garcez, C.C.R.; de Lima, D.V.; Miranda, R.K.; Mendonça, F.; da Costa, J.P.C.L.; de Almeida, A.L.F.; de Sousa, R.T., Jr. Tensor-Based Subspace Tracking for Time-Delay Estimation in GNSS Multi-Antenna Receivers. *Sensors* **2019**, *19*, 5076. doi:10.3390/s19235076. [CrossRef]
13. Xie, T.; Jiang, H.; Zhao, X.; Zhang, C. A Wi-Fi-Based Wireless Indoor Position Sensing System with Multipath Interference Mitigation. *Sensors* **2019**, *19*, 3983. doi:10.3390/s19183983. [CrossRef] [PubMed]
14. Schmidhammer, M.; Gentner, C.; Siebler, B.; Sand, S. Localization and Tracking of Discrete Mobile Scatterers in Vehicular Environments Using Delay Estimates. *Sensors* **2019**, *19*, 4802. doi:10.3390/s19214802. [CrossRef] [PubMed]
15. Emadzadeh, A.A.; Lopes, C.G.; Speyer, J.L. Online time delay estimation of pulsar signals for relative navigation using adaptive filters. In Proceedings of the 2008 IEEE/ION Position, Location and Navigation Symposium, Monterey, CA, USA, 5–8 May 2008; pp. 714–719.
16. Liu, X.; Yang, X.; Xiong, W. A robust global approach for LPV FIR model identification with time-varying time delays. *J. Franklin Inst.* **2018**, *355*, 7401–7416. doi:10.1016/j.jfranklin.2018.07.025. [CrossRef]
17. Park, P.; Di Marco, P.; Shin, H.; Bang, J. Fault Detection and Diagnosis Using Combined Autoencoder and Long Short-Term Memory Network. *Sensors* **2019**, *19*, 4612. doi:10.3390/s19214612. [CrossRef] [PubMed]
18. Srinivasa Rao, D.; Siva Kumar, M.; Ramalinga Raju, M. Enhancement of robust performance for fuzzy parametric uncertain time-delay systems using differential evolution algorithm. *Int. J. Robust Nonlinear Control* **2019**, *29*, 6337–6356. doi:10.1002/rnc.4694. [CrossRef]
19. Shen, S.; Song, A.; Li, T.; Li, H. Time delay compensation for nonlinear bilateral teleoperation: A motion prediction approach. *Trans. Inst. Meas. Control* **2019**, *41*, 4488–4498. doi:10.1177/0142331219860928. [CrossRef]
20. You, F.; Cheng, S.; Zhang, X.; Chen, N. Robust fault estimation for Takagi-Sugeno fuzzy systems with state time-varying delay. *Int. J. Adapt Control Signal Process.* **2019**, doi:10.1002/acs.3073. [CrossRef]
21. Pfeiffer, J.; Wu, X. Automated Time Delay Estimation for Distributed Sensor Systems of Electric Vehicles. In Proceedings of the 2019 10th IEEE International Conference on Intelligent Data Acquisition and Advanced Computing Systems: Technology and Applications (IDAACS), Metz, France, 18–21 September 2019; pp. 609–614.
22. Barber, D. *Bayesian Reasoning and Machine Learning*; Cambridge University Press: Cambridge, UK, 2011.

sensors

MDPI

Article

Advanced UAV–WSN System for Intelligent Monitoring in Precision Agriculture †

Dan Popescu *, Florin Stoican, Grigore Stamatescu, Loretta Ichim and Cristian Dragana

Faculty of Automatic Control and Computers, University Politehnica of Bucharest, 060042 Bucharest, Romania; florin.stoican@upb.ro (F.S.); grigore.stamatescu@upb.ro (G.S.); iloretta@yahoo.com (L.I.); cristian.dragana@gmail.com (C.D.)

* Correspondence: dan_popescu_2002@yahoo.com; Tel.: +40-766-218-363

† This paper is a considerably extended version of our paper published in Dan, P.; Florin, S.; Loretta, I.; Grigore, S.; Cristian, D. Collaborative UAV-WSN system for data acquisition and processing in agriculture. In Proceedings of the 10th IEEE International Conference on Intelligent Data Acquisition and Advanced Computing Systems: Technology and Applications (IDAACS 2019), Metz, France, 18–21 September 2019.

Received: 31 December 2019; Accepted: 31 January 2020; Published: 3 February 2020

check for updates

Abstract: The growing need for food worldwide requires the development of a high-performance, high-productivity, and sustainable agriculture, which implies the introduction of new technologies into monitoring activities related to control and decision-making. In this regard, this paper presents a hierarchical structure based on the collaboration between unmanned aerial vehicles (UAVs) and federated wireless sensor networks (WSNs) for crop monitoring in precision agriculture. The integration of UAVs with intelligent, ground WSNs, and IoT proved to be a robust and efficient solution for data collection, control, analysis, and decisions in such specialized applications. Key advantages lay in online data collection and relaying to a central monitoring point, while effectively managing network load and latency through optimized UAV trajectories and in situ data processing. Two important aspects of the collaboration were considered: designing the UAV trajectories for efficient data collection and implementing effective data processing algorithms (consensus and symbolic aggregate approximation) at the network level for the transmission of the relevant data. The experiments were carried out at a Romanian research institute where different crops and methods are developed. The results demonstrate that the collaborative UAV–WSN–IoT approach increases the performances in both precision agriculture and ecological agriculture.

Keywords: unmanned aerial vehicles; wireless sensor networks; intelligent data processing; trajectory planning; relevant data extraction; data consensus; Internet of Things; precision agriculture

1. Introduction

The need for high-performance, high-productivity, and sustainable agriculture results from the rapid growth of the human population. This requires permanent monitoring and intelligent processing of the measured data collected from the field, correlated with the weather forecasts, to produce agronomic recommendations. In the last few years, new technologies in agriculture, and especially in precision agriculture (PA), have been leveraged for increased productivity and efficient input dosage [1]. Most importantly, in PA, farmers need to know exact and timely details about crop status. These details about certain parameters, obtained by measurements both from the ground and in the air, constitute input data to specialized systems of process management in the PA. Some relevant examples might include for example, irrigation control, pesticide dosage, pest control, etc. For acquisition and complex processing of the collected data, integration of unmanned aerial vehicles (UAV) with wireless sensor networks (WSN) under novel frameworks, such as the Internet of Things (IoT), has been shown to contribute to increases in agricultural yields [2]. Such advanced systems are modeled as well-specified agent-based solutions with sensors and UAVs. Although the contributions of UAVs and WSNs, taken separately, are well documented and important in the sustainable growth of agricultural production, the integration of these components together within an IoT framework is expected to significantly improve the solutions for monitoring, production modeling, prediction, and decision-making.

Relevant applications of UAV–WSN systems are presented in Reference [3–6]. In viticulture, as a special type of PA, the soil and air parameters modify grape yield and quality. For this purpose, a solution based on the collaborative system mini UAV (quadrotor type)–WSNs to monitor parameters, like temperature and humidity, to prevent the frost in fragmented vineyards is proposed in Reference [3]. The UAV is considered as communication relay between sensors and a base station. A real application for monitoring sensitive parameters in vineyards with both agro-meteorological stations and UAV platforms is presented in Reference [4]. In order to obtain a precise monitoring of the specific indicators, the data from the ground are correlated with the data collected by a UAV platform with 8 rotors provided with a professional thermal camera. The study was conducted over a period of two years. For data acquisition on large areas, a fixed wing type UAV, used as data mule from the ground WSN, is proposed in Reference [5]. In addition, the UAV has attached an high definition (HD) camera for the detection of certain plant diseases. Experimentally, a small tank has been added to spray different insecticides, fertilizers, herbicides, etc. Both UAV and WSN are low cost and not robust for only demonstration purposes. In addition, in Reference [6], a low cost agro-meteorological monitoring system in a vineyard was designed and developed. The optimal positioning of the sensors was made with the help of the multispectral image analysis, acquired by UAV.

Given recent evolutions in UAV technologies, cost reduction, and new regulations of aviation authorities regarding the usage and deployment of such systems (e.g., European Aviation Safety Agency (EASA) [7] and Federal Aviation Administration (FAA) [8]), such aerial robotic platforms are increasingly used in agriculture for different tasks, the most important being crop monitoring [9]. According to EASA, the UAVs should be safely integrated into the existing aviation context in a proportionate way [7]. For large scale applications, in which UAVs are flying beyond line-of-sight, compliance with strict regulatory frameworks is essential.

Adoption of a UAV-based solution for image acquisition in agriculture applications is more cost effective and flexible in comparison with satellite or manned aircraft alternatives [10]. Both fixed- wing [11,12] and rotary-wing type [3,13] UAVs are frequently used in various applications in agriculture, while accounting for the risk of crashes [9] and potential damages. Equipped with specific sensors in modular payloads [14], such as high resolution RGB [15], infrared, multispectral [16,17], thermal cameras [18,19], and also LIDAR [10], UAVs are able to create precise maps of crop state or evolution [17], health plant assessment [20], diseases [21], soil characteristics, evaluate losses caused by floods [11], etc. In the crop monitoring, the following characteristics are analyzed from UAV images [9]: the crop water stress, defined as the difference between the canopy and the air temperature, the photochemical reflectance index, and the vegetation indices.

Although UAVs with different propulsion systems are now available, most applications in PA use UAVs driven by electric motors due to their compact size, reduced maintenance and operational costs and, not the least, their alignment with the current regulatory context and tendencies towards the reduction of global carbon emissions [22].

The small-scale data acquisition by the WSN helps farmers to take actions like crop irrigation, fertilizer usages, deciding on the optimum stages of sowing, and harvesting [23]. Moreover, WSNs employed in PA lead to large amounts of data. Thus, data collection by WSNs is an important contribution to the development of farm management information systems (FMIS) [24,25].

The WSN has multiple functions at the field level: data acquisition of various parameters (e.g., temperature in soil and air, humidity in soil and air, solar radiance, soil nutrients, the presence of pests and weeds, chlorophyll content in plants, etc.), distributed processing of data by establishing consensus—if it is the case, establishing the relevant data and its storage, low level data fusion, and data transmission. New sensor node designs offer reduced costs [26]; see, for example, the detailed list of sensors used in PA given in Reference [10]. As in many other large-area monitoring applications, for communication or local processing reasons, the sensors are grouped into sensor networks, the communication being made by radio. A WSN network will include measurement nodes (sensory nodes) and data collection, processing, and transmission nodes (sink or cluster head).

Regarding PA, there is no rigorous theory of sensor placement because it depends on the particularities of the soil and the weather. Sensor groups need to comply broadly with the need for sensory and communication coverage. In Reference [27], two examples of sensor location topologies are given: grid and random. From the point of view of communication with the sink node, the most used are the star and mesh topologies. The wireless communication protocols used in WSN for PA are the following [10]: 6LoWPAN, ZigBee (both being the most suitable for the mesh topology), LoRaWAN, GSM, BLE, and Wi-Fi.

In PA, WSNs are used, most often, for parameter monitoring, but they can also be integrated into control systems as sensors. Direct specific applications of WSN in control systems for PA are the energy efficient automated control of irrigation [28] and smart automated fertilization [29].

The performance of the crop monitoring can be improved by UAV–WSN collaboration [30]. The collaborative aspects in an integrated UAV (aerial agents)–WSN (ground agents) architecture for different applications was recently presented in a review paper [22], where the different functional components of the system and how they collaborate with each other was highlighted. In Reference [31], the authors presented an integrated UAV–WSN–IoT system, named FarmBeats, which is an end-to-end platform for data collection from various sensors, cameras, and drones in agricultural applications. An unlicensed TV White Spaces is used to setup a high bandwidth link from the farmer's home to an IoT ground station at a distance for collecting data from UAVs and WSNs.

In order to interconnect the UAVs and terrestrial WSNs into hybrid networks and, at the same time, to ensure a safe airspace sharing with aircrafts, multiple organizations are contributing [22]: International Civil Aviation Organization, EASA, Joint Authorities for Rulemaking on Unmanned Systems, International Telecommunications Union, etc. Satellite connection is required for two reasons. One-way communication, such as obtaining the GPS location of the UAVs or the sensory nodes (if any) is one reason. The second reason is a possible data transmission or remote control (via two-way satellite-intermediated internet).

In Reference [32], the authors discuss the information system design supporting agriculture data management. Enabling advanced data processing in the form of sensor fusion and clustering mechanisms for improved network topologies in generic applications has been discussed [30]. Effective data gathering mechanisms [33] and higher level IoT architectures [34] are key and current topics of interest.

We believe that the challenges of UAV–WSN–IoT integrated systems can come from several directions: (a) precise localization of the ground sensors with the aid of a preliminary flight; (b) sensor states periodically inspected by UAV; (c) establishing of the WSNs as sensor clusters able to cover, both from the sensorial and from the communication point of view the monitored area; (d) establishing the cluster heads (CH), named base stations, of the WSNs able to communicate data to UAVs; (e) transmitting commands to change the strategy and parameters of the sensor networks, (f) data acquisition from WSNs through UAVs, (g) special trajectory planning and tracking, (h) the aggregation of information collected by the UAV with the information collected by WSN for the purpose of measuring and interpreting the parameters with increased accuracy, (i) remote control via Internet, and (j) edge and cloud computing.

In a hierarchical structure, the data processing architecture of the integrated system is based on three levels: consensus, edge computing [35], and cloud computing.

For the main activity, the data collection from CH, UAV must have a predefined trajectory, properly designed, and accounting for the following limitations:

- Waypoint passing: a UAV has to pass above the CH to extract the relevant data from that area (covered by the corresponding WSN sub-network);

- Obstacle avoidance: UAVs avoid obstructions or prohibited areas along the flight plan;
- Guaranteed communication: to ensure that the data has been fully collected, enough time has to be spent in the CH neighborhood;
- Efficiency: reduce at a minimum the energy consumption for that trajectory (consider the length of the trajectory and its complexity).

The integration of UAV–WSN based systems for PA in IoT is a mandatory step to create an advanced FMIS [25].

Due to the integration, the system can become "smart" by using elements of artificial intelligence like self-adaptation and decision, optimal trajectory, data transmission of relevant parameter values, energy efficiency, and neural networks for data and image processing. Not in the least, the sensors must be placed optimally, considering the terrain characteristics. Battery life is an important design point of the ground sensor algorithms by reducing to a minimum the number of wireless communications needed to transfer the information. The radio interface is the critical factor in increasing battery life. Based on the frequency of the data collection and radio transmissions the nodes can have a battery lifetime ranging from several months up to one year. Therefore, the intelligent collaboration between UAV and WSN can lead to optimization of parameters, such as energy consumption, sensing coverage, risk, data acquisition, and processing time [36]. To this end, bio-inspired optimization heuristics and genetic algorithms were applied to the aforementioned agents.

The optimal WSN coverage by the aid of UAV platforms is implemented in Reference [37] as an optimization problem, formulated by means of the travelling salesman problem, in order to find the best path of the UAV for data collection with minimum energy consumption.

Using UAV as data mule for multi WSNs is an energy-efficient method to increase the networks' life. To this end, the authors in Reference [38] apply the successive convex optimization technique.

The proposed system presents the following integration aspects:

- Group the sensors in clusters and determine the cluster heads, the methodology proposed by the authors in Reference [30];
- Path planning based on specific conditions for efficient data collection; and
- Intelligent data collection and processing.

The main contributions consist in the following: (i) implementation of a multilevel, collaborative UAV–WSN system structure for agriculture applications, (ii) a specific path planning for fixed wing–type UAV with the purpose of robust and efficient data collection, (iii) obtaining relevant data from sensors for the purpose of saving energy, and (iv) edge–fog–cloud computing algorithms for subsequent data processing. Thus, the main challenge is related to improving data extraction and communication in large scale heterogeneous monitoring system. The key problem is focused on improving the performance of such systems through better algorithms and synchronization among the two subsystems: the ground sensor network and the robotic aerial platforms, implemented as UAVs, for data collection and relaying.

The rest of the paper is structured as follows. Section 2 describes the concept, the methodology, and key aspects that have been addressed for the proper design and implementation of the system. Section 3 presents the experimental results and performances after implementing the system on an experimental farm. Section 4 highlights the conclusions, as well as future work.

2. Materials and Methods

2.1. *Requirements for Integrated UAV-WSN-IoT Systems*

For the design of reliable and robust large-scale monitoring system the requirements have to first be validated. The main challenges for such collaborative systems were considered to be: sensing coverage in accordance to mission objectives, communication coverage by the hybrid UAV–WSN system using various types of radio links, from low-power, low-data rate to high throughput long distance for streaming, energy efficiency, and, not in the least, computing efficiency. The decentralized architecture for crop field monitoring described in this paper is designed to overcome the challenges mentioned above and to account for the data generation patterns at the field level. While the proposed data fusion mechanisms and processing of centralized in-field data at CH level manage to reduce data volume and ensure the flow of information up to the level of events, an additional intermediate level is appended to the data stream, in order to reach the server. To this end, we consider both mobile agents (UAV) and multiple fixed agents (ground sensors (SNs)). The system diagram is presented in Figure 1. The mobile agent can perform the following functions: data mulling, image acquisition, relay, and state inspection of WSNs. The fixed agents acquire data from the field (agricultural field—soil and air), process data locally (relevant data extraction, data consensus), and finally transmit data to the UAV by means of CHs. The system is composed of four main processing levels (Table 1): Sensor level, Fog Computing level, Internet/Cloud Computing level, and Data Management and Interpretation level. This is a multi-WSN–UAV structure with higher level integration in internet-based systems for decision support. The data from WSNs are collected by a UAV, transmitted at a ground control station (GCS), and, from here via the internet, to the Data Interpretation module. Analytics functionality ranges from basic statistical indicators to trend and event detectors and up to basic statistical learning models that have the ability to anticipate evolutions in the monitored ground phenomena.

Another important requirement of the integrated system is the correlated or complementary interpretation of the data from the sensory agents, either mobile or fixed. For example, when the soil moisture is too high, the soil sensors show the maximum value and cannot discern whether a flood has occurred. This can be accurately determined from aerial images taken by the UAV. In addition, the degree of humidity in plants and the degree of foliage development can be observed either from the ground or from the air (images), and a more precise determination results from the fusion of the two data sets.

Other types of similar systems were surveyed and can include the use of swarms of multi-copter type UAVs, which offer better positioning accuracy for data collection while trading off energy efficiency and autonomy. Ground sensor network implementation can also be a differentiating factor with two main approaches: random deployment of sensor nodes in the area of interest, according to a minimum expected sensing coverage density, or deterministic, grid-like placement. Intermediate data processing steps from the field level to the decision level are commonly accepted as an important mechanism to balance network loads and improve communication latency.

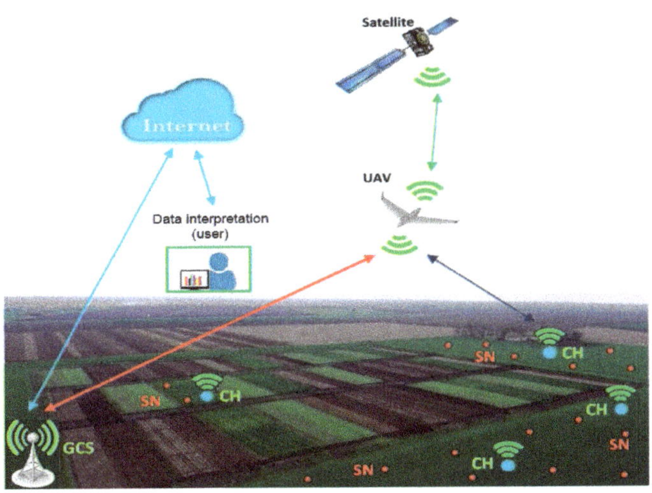

Figure 1. The concept of the integrated unmanned aerial vehicle (UAV)–wireless sensor network (WSN)–Internet of Things (IoT) system.

Table 1. Processing levels.

Level	Content
Field	Sensors (SNs)
Edge computing	Cluster heads (CHs), UAV
Cloud computing	Cloud
Data interpretation	User server

2.2. UAV Trajectory Design

For UAV trajectory planning, two cases must be considered. The first is the trajectory planning for collecting data from sensors (CHs), and it must take into account certain requirements, such as the complete and safe acquisition of data, on one hand, and minimize energy and time consumption, on the other hand [39].

Under certain reasonable assumptions (known environment, known limitations), the UAV tasks reduce to computing a trajectory which respects constraints and minimizes a cost (length, total energy consumption, etc.), while simultaneously respecting various constraints (internal dynamics, stall velocity constraints or exogenous ones, those imposed by the environment, such as obstacle avoidance and waypoint passing through).

The particularity lies in the fact that many of the UAV-specific constraints are non-convex [40], e.g., the variable of interest z (depending of time t) has to stay outside some bound (e.g., outside of an interdicted region and/or maintain a minimal velocity). If $z(t)$ is the UAV position, the velocity restrictions are usually written as follows:

$$\underline{v} \leq \left\|\dot{z}(t)\right\| \leq \overline{v}. \tag{1}$$

Both bounds (lower—$\underline{v}$ and upper—$\overline{v}$) may depend on a variety of factors. Hard constraints are imposed by the UAV physics: upper bound given by the engine characteristics and lower bound by the requirement to avoid stall. Note that this work neglects the influence of wind: velocity is usually measured against the ground (e.g., through a GPS), but, in fact, the UAV "feels" the addition of its own and of the wind velocities. This may lead to an unexpected stall or, at least, improper behavior. Usual techniques are to provide more conservative bounds in Equation (1) and to restrict the flight to normal weather conditions.

Waypoints are introduced, in a practical mission, because data has to be gathered from a cluster node. Thus, the question of minimum communication time arises [41]: It is necessary to remain in a specific neighborhood for a defined time interval Δt_i. To correctly describe such a constraint, we require a tuple $(\omega_i, \Delta t_i, r_i, R_i)$, where ω_i is the corresponding cluster node position (the center of the circle in Figure 2), and r_i and R_i are, respectively, the minimum and the maximum radius of the permitted communication area. Because there are perturbations due to trajectory control errors or other causes, the real trajectory is included in a flight lane (Figure 2a). The flight lane was experimentally established at 30 m, under reasonable assumptions about wind speed. The trajectory $z(t)$ has to stay near the waypoint for a least amount of time Δt_i determined by the quantity of data which has to be transferred:

$$r_i \leq \left\|\omega_i - z(t)\right\| \leq R_i,\ t \in [t_i, t_i + \Delta t_i]. \tag{2}$$

Condition (2) is often impractical to check due to the continuous nature of $z(t)$ and because of the varying time interval $[t_i, t_i + \Delta t_i]$. The usual approach is to sample the constraint and to estimate the path length by assuming the bounds (1) on the velocity. To this end, we consider:

$$\left\|z(\bar{t}_i) - \omega_i\right\| = r_i, \tag{3}$$

with $\bar{t}_i$ given such that $\bar{t}_i \in [t_i, t_i + \Delta t_i])$ holds; it is important that a waypoint is reached, not when.

Note that the shortest distance for a trajectory checking (4) is the straight line shown in Figure 2a, whose length is $2\sqrt{R_i^2 - r_i^2}$. In other words, a sufficient condition for guaranteeing that the minimal time Δt_i has passed is to ensure that

$$\Delta t_i \geq \frac{2\sqrt{R_i^2 - r_i^2}}{\overline{v}}. \tag{4}$$

Condition (4) provides a lower bound for the time the UAV stays between the inner and outer circles (i.e., how much time it spends inside waypoint's ω_i communication range). Then, inserting (3) in a trajectory design procedure implicitly guarantees enough communication time. This approach may be insufficient for a couple of reasons. First, the desired communication time may not be known at the trajectory generation time and thus could not be compared with Δt_i. Second, the communication time is known to be larger than Δt_i and a "tangential" pass (like the one enforced by (3)) does not suffice. The method (detailed below) is to enter a loitering mode to increase arbitrarily the data-gathering time [42]. Making the reasonable assumption that the loitering r_i^l radius respects the condition $r_i < r_i^l < R_i$, means that the UAV can orbit the waypoint ω_i for an indefinite period of time [43]. From the viewpoint of trajectory generation, the only relevant question remains the places at which the UAV inserts/dislodges onto/from the loitering circle. Both of these points are decided by the relative position of the current waypoint with respect to the previous and next waypoints in the sequence (such as to reduce unnecessary inflexions in the trajectory). The switch between normal and loitering modes will be done at pre-determined points: the trajectory enters loitering mode at a point ω_i^- and dislodges from it at a point ω_i^+ (which lie on the loitering circle and are from/towards the direction of the previous/next waypoint). Thus, when the UAV decides to finish the communication, it will continue to orbit the loitering circle until it reaches the dislodging point ω_i^+. Here, it will switch back to the normal trajectory mode.

The inner (dotted line), outer (solid line) communication circles, and loitering circle (dashed line) are illustrated in Figure 2b. We show a trajectory inserting to the loitering circle, tracking an arc of it, and, lastly, dislodging from the circle to re-enter its normal mode (line tracking). The UAV could have orbited the loitering circle repeatedly and dislodged from it at ω_i^+ when desired. As is mentioned above, the trajectory describes a corridor (we account for the inherent tracking error appearing under realistic conditions).

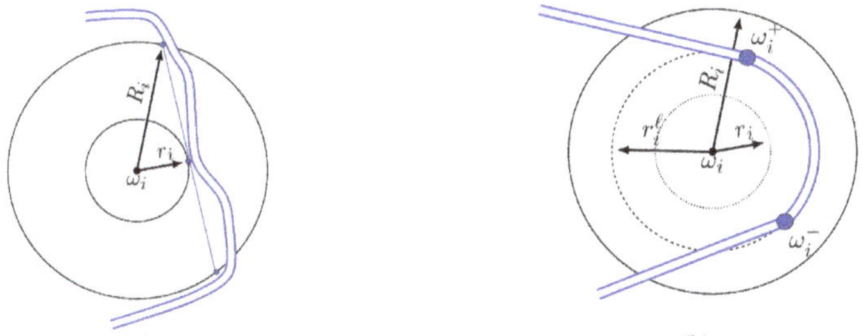

(a) **(b)**

Figure 2. Illustration of different aspects of the trajectory design: (**a**) inner and outer communication constraints with a sufficient condition and a corridor for the UAV trajectory envelope and (**b**) trajectory validating.

While the previous velocity and time constraints are easy to formulate, they lead to complex (nonlinear in position and time variables) constraints. Thus, in practical implementations, it is often much easier to provide a simplified control scheme based on the heading angle (a "line of sight" procedure).

That is, the UAV control is partitioned into the lower level where the velocity is controlled (to negate the wind disturbances, for example) and the higher level where, at each time instant, a new heading angle is computed. Thus, we may interpret the path as a collection of segments (linking consecutive waypoints) and circle arcs around waypoints where loitering is needed.

The idea of the segment tracking procedure is straightforward and is sketched in the following flowchart (Figure 3). In the flowchart, we make use of several notations:

- RTB = return to base, a flag denoting whether the UAV has to return to its path's starting point;
- LM = loiter mode, denotes that the UAV has entered the loiter mode; at the start of this mode, the LMT = loiter mode remaining time is initialized to a predefined value which is decreased (at each step with a constant value T) as long as the UAV remains in the loiter mode;
- PP = projection point, obtained by projecting the current position onto the support line of the current segment from which W = weight of the PP (denoting whether the PP is inside the segment, to the left or to the right) and D = distance between the UAV position and the PP, are computed;
- PCP = proximity circle point represents the intersection between the proximity circle and the current segment (in case of intersection between the circle and the segment there are two solutions; the one closest to the end-point of the segment is taken);
- LP = loiter point is computed such that the UAV tracks the loiter circle (with the sense of movement decided a priori by the supervisor); and
- CP = current waypoint, throughout the algorithm, is updated as needed.

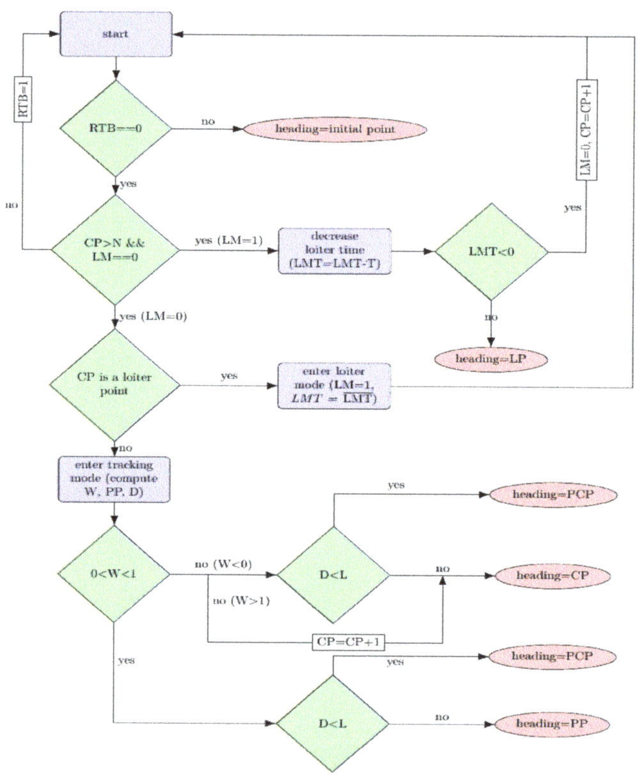

Figure 3. Flowchart for the path planning. RTB = return to base; CP = current waypoint; LM = loiter mode; LMT = loiter mode remaining time; T = constant value; LP = loiter point; PP = projection point; W = weight of the PP; D = distance between the UAV position and the PP; PCP = proximity circle point; L = vector length.

The main points of the algorithm are:

- The UAV has two modes of functioning, loiter mode and segment tracking mode, which are decided by the supervisor (in the sense that within the collection of waypoints a priori computed, some of them are labeled as loiter points).
- In both cases, the algorithm provides a heading which is the reference to be tracked by the UAV. This is in line with standard practices, where the heading is decided through some design procedure and the velocity and pitch and roll angles are decided at the auto-pilot level (usually the velocity is maintained constant and the roll and pitch are taken as needed between admissible bounds).
- The decisions taken by the algorithm and supervisor are, ultimately, related to the distance between the current position and some point of interest. To do so, we consider some circles of interest, defined as follows:
    - Communication circle: the UAV communicates with the ground-based cluster head only when it is within the communication radius.
    - Waypoint update circle: it is impractical to assume that the UAV passes through the exact coordinates of the current waypoint. Thus, we update the active segment (by advancing through the list of waypoints) whenever we are close enough to the end-point of the current segment.
    - Loitering circle: whenever the UAV is required to spend a significant time in communication with the current cluster head, the decision to start loitering is taken. The loitering radius is restricted to be less than the communication radius and larger than the physical limitations imposed by the roll angle bounds (a tighter circle means a larger roll angle).
    - Proximity circle: the procedure employed in the algorithm takes (whenever there is intersection between the circle and the current segment) the heading angle in the direction of the intersection point (the one closest to the end-point of the segment).
- When the last waypoint is covered, the UAV returns to base (by default, we consider this to be the initial point on the trajectory).

Without being exhaustive, some of the most relevant updates in the algorithm are:
In segment tracking mode:

1. At the current time, we consider the UAV position (x,y), the segment determined by the current (CP) and next waypoint (CP + 1): $\left(w_x^i, w_y^i\right)$, $\left(w_x^{i+1}, w_y^{i+1}\right)$.
2. We compute the projection of the current point onto the current segment (PP). We identify three possible cases by checking the relative position of the projection wrt the segment's end points (described by W): inside the segment ($0 \leq W \leq 1$)), outside and located before the initial segment end ($W < 0$); outside and located after the initial segment end ($W > 1$);
3. We compute the distance (D) from the current point to the segment and the circle of radius L (proportional with the UAV velocity) and further used to compute the heading vector.
4. We consider the following cases:
    i. The UAV is too far away, and the projection point lies before the segment start point. Then, the heading angle points towards the projection point.
    ii. The UAV is sufficiently close, and the projection point lies before the segment start point. Then, the heading angle points towards the start point.
    iii. The UAV is sufficiently close to the segment end point, or its projection onto the segment lies after the end point. Then, the current segment is updated, and the procedure jumps to step 4.i.

 iv. The UAV is too far away, and its projection lies onto the interior of the segment. Then, the heading vector points towards the projection.
 v. The UAV is sufficiently close, and its projection lies onto the interior of the segment. The heading angle is taken as the vector of length L in which the tip lies on the segment (there are two possible tips; the one closer to the segment end point is considered).

5. Go to step 1.

In the loitering mode:

1. Select the loitering center as the current waypoint.
2. Construct the circle of radius L and centered in the current position of the UAV.
3. If the circle does not intersect the loitering circle, move towards the projection point situated on the loitering circle.
4. If the proximity circle intersects the loitering circle, take the heading vector along the tangent at the intersection point between loitering circle and proximity circle (there are two solutions, we selected depending on the desired loitering rotation—clockwise or counterclockwise).

Note that all steps where a decision regarding the trajectory update is taken consist in fact in a decision about the UAV's heading. Thus, for trajectory tracking, only the heading angle is used as control input. This suffices for relatively simple trajectories and is robust against wind disturbances (as later shown in the simulations).

2.3. Relevant Data Extraction

The collected data is hierarchically processed from the ground level, cluster head level, UAV level up to the cloud. Alongside these steps, information is gradually extracted through various methods that enable local decisions based on the configuration of the system (thresholding, consensus, symbolic aggregate approximation, etc.).

In-field data processing is ensured both at local level (independent data filtering) and decentralized at network level (through data exchange between neighbor sensory nodes). The proposed data processing mechanisms, tailored for in-field level, are designed in order to ensure a substantial weighted average. This step is found as 'Enable consensus dialog'. Once the convergence is reached, each node performs a routine for results analysis basically seeking to discover and mark nodes with divergent values. This information remains available alongside the consensus value so that it can be interrogated by the higher level of data processing if needed. This is found in Figure 4 as 'Analyze results step'.

Aggregated data sets are achieved through different methods. All seek for relevant data points, aiming to a reduced size set and providing at the same time a satisfying reconstruction of the initial data. The proposed method for data aggregation is based on the minimum and maximum values extraction, computed as global extremes for a predefined period of time (e.g., a day). It is obvious that this method is suitable only for measurements that have a periodic behavior, with smooth variations during the day. A measurement for which this method is suitable is the soil temperature. Conversely, change detection is commonly used for irregularly-shaped data sets. This method follows extraction of local extreme points where trend changes occur.

Given a set of data points (x_i, y_i), $i = 1, \ldots, n$, trend t_i is computed for each sequence measurements such that for a measure m, (5),(6),(7) has to be true. If $t_i \neq t_{i+1}$, then it means that a trend change has occurred, and the data point (x_i, y_i) is added to the relevant data set.

$$x_{i+1}^m - x_i^m > \delta^m \Rightarrow t_i^m = 1, \tag{5}$$

$$x_{i+1}^m - x_i^m < -\delta^m \Rightarrow t_i^m = -1, \tag{6}$$

$$x_{i+1}^m - x_i^m \in [-\delta^m, \delta^m] \Rightarrow t_i^m = 0. \tag{7}$$

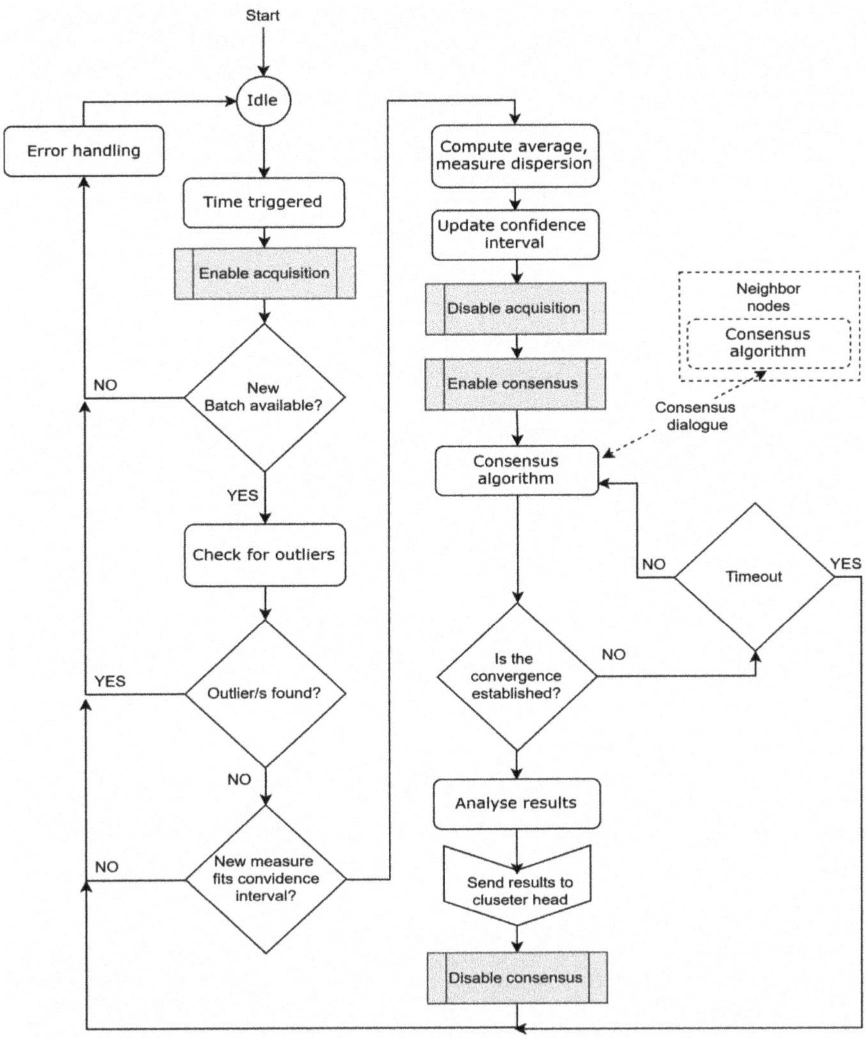

Figure 4. Flow diagram of the data processing steps at the field level, based on consensus algorithm.

Data collection is done periodically, following a succession of specific routines. As mentioned before, the first step for in-field data processing is performed at the local level, independently, by each sensor node.

While the proposed data fusion mechanisms and processing of centralized in-field data at gateway level manage to reduce data volume and ensure the flow of information up to the level of events, an additional intermediate level is appended on the data stream, in order to reach the server. Consequently, the system is composed from three processing levels (Figure 5): In-field data processing, Edge computing, and Cloud computing. This corresponds to a UAV–WSN system with internet integration. The data from WSNs are collected by a UAV (or team of UAVs) and then transmitted at a ground control station (GCS). From here, the data is transmitted, via the internet, to the Cloud computing level and, finally, to the 'Data interpretation and decision' module.

In a consensus mechanism, multiple autonomous agents seek to reach the convergence value under the influence of the information flow exchanged inside the network. Each node updates its estimated value using an updating rule. An update law for node n_i based on local weighted consensus is described by the following equation:

$$x_i(k+1) = \omega_{ii} x_i(k) + \sum_{j \in N_i} \omega_{ij} x_j(k), \tag{8}$$

$$\sum_{i \in M} \sum_{j \in N_i} \omega_{ij} = 1, \tag{9}$$

where

$x_i \in \mathbb{R}$ is the computed estimate of node i;
ω_{ii} is the weight applied to its own previous computed estimate;
ω_{ij} is the weight associated with the node j for the value of node i;
k is a convergence step; and
N_i is the neighborhood of node i, $i \in \{1, 2, 3, \ldots, m\} = M$.

The proposed consensus algorithm is built using a hybrid weighted average consensus which ensures that the updating rule computes the current convergence value, keeping a high priority for the closest neighbors, but at the same time, it aims at suppressing outlier values.

Each node computes the weights ω_{ij} based on the distance d_{ij} computed using the available location information.

$$\omega_{ij} = \begin{cases} \frac{d_{min}}{d_{ij}} & if\ (i,j) \in \varepsilon,\ i \neq j \\ 0 & if\ (i,j) \notin \varepsilon,\ i \neq j \end{cases}, \tag{10}$$

where

d_{min} is the distance to the closest neighbor; and
d_{ij} denotes the distance between node i and j.

Using the selected weights, the algorithm performs a weighted average of neighbors values defined as:

$$N_i mean(k+1) = \frac{\sum_{j \in N_i} \omega_{ij} x_j(k)}{dim(N_i)}. \tag{11}$$

In order to suppress outlier values, additional weights are applied for previously computed estimate $x_i(k)$ and current neighborhood estimate average $N_i mean(k+1)$. Thus, this is an auto-supressing mechanism computed as the ratio between the standard deviation at convergence step $k+1$ and the deviation of the previous estimate $x_i(k)$. This is written as:

$$\begin{gathered} x_i(k+1) = \frac{\Delta(k+1) \cdot x_i(k) + \delta \cdot N_i mean(k+1)}{\Delta(k+1) + \delta} \\ \Delta(k+1) = \frac{\sqrt{\frac{\sum_{j \in N_i} \left[x_j(\mathrm{k}) - N_i mean(k+1) \right]^2}{N_i - 1}}}{\sqrt{\left[x_i(\mathrm{k}) - N_i mean(k+1) \right]^2}}, \\ \delta = 1 - \Delta(k+1) \end{gathered} \tag{12}$$

where

- $\Delta(k+1)$ is the weight applied to the state value, computed for each step of the average consensus;
- δ is the weight applied to the neighborhood estimate.

Once the consensus is reached, each node performs a routine for results analysis basically seeking to discover and mark nodes with divergent values. This information remains available alongside the consensus value so that it can be interrogated by the higher level of data processing if needed.

This global mechanism indicates problematic sensor nodes or even very isolated events, but it cannot discern between them.

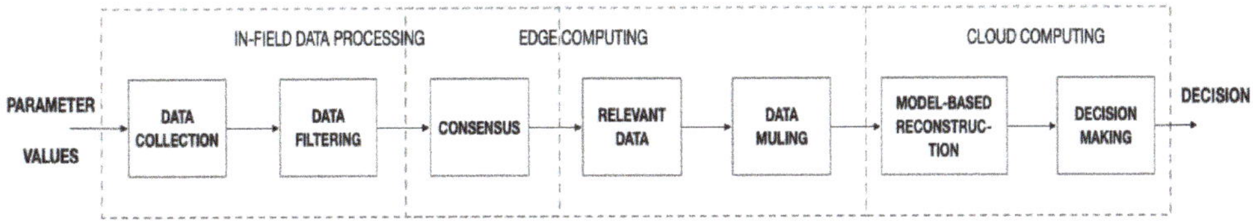

Figure 5. Flow diagram of the data processing at the system level.

The flow diagram presented in Figure 5 shows the data processing pipeline for the integrated UAV–WSN–IoT system. Based on preliminary parameterization, e.g., sample rate, coverage area, and energy aware communication, sensor measurements are collected at the ground level by the local nodes. On-board basic data filtering is carried out to check the consistency and validity of the measurements for early detection of sensor faults, misreading or outliers. At the local network level, based on the validated and filtered data, consensus-based agreement is performed by in-network data processing, which leads to a common value for each of the acquired parameters among all nodes in a cluster. The cluster head further operates on the data by extracting relevant information through edge computing mechanisms, and a model-based compressed representation is achieved, e.g., polynomial interpolation models or more advanced methods, such as SAX (Symbolic Aggregate Approximation). At the conclusion of the edge computing phase, the UAV is activated for collecting the compressed representations of the ground phenomena from the cluster head nodes. The trajectory of the UAV is optimized as previously discussed to ensure timely collection from all the cluster heads in a target area and transfer the data to a central unit for back-end cloud computing processing and decision. The cloud computing layer integrates the data reconstruction based on the model parameters as inputs to a decision-making process, which yields the final outcome and allows for closing the loop by acting on the ground environment, e.g., irrigation and input dosage signals for the precision agriculture application.

When it comes to processing a large volume of data, many high-level representations of time series have been proposed for data mining, including Fourier transforms, wavelets, and piecewise polynomial models [44]. A different approach that we consider is the SAX algorithm, proposed in Reference [45]. This is a flexible method that allows adjusting the ratio between data volume and data relevance to ensure a fair reconstruction of original trends, while ensuring high data reduction by transforming of a time series into text strings. In essence, the algorithm operates by assigning label symbols to segments of the time series, thus porting it in a unified lower dimension representation. The importance of SAX′ parameterization must be considered by defining the number of segments and the alphabet size.

Starting with a time series X of length n, this is approximated into a vector $\overline{X} = (\overline{x}_1, \ldots, \overline{x}_M)$ of any length $M \leq n$, with n divisible by M. Each element of the vector $\overline{x_i}$ is calculated by:

$$\overline{x_i} = \frac{M}{n} \sum_{j=\frac{n}{M(i-1)}+1}^{(n/M)_i} x_j. \quad (13)$$

3. Experimental Results

The high-level configuration of the integrated system is illustrated in Figure 6. The UAV is of the fixed wing-type, which enables coverage of large geographic areas with low energy consumption. The base station (CH) collects the primary data processed from the field sensors and periodically transmits it to a UAV according to its synchronization with the planned trajectory. Further, the data are processed in the cloud after the UAV uploads the collected data over the internet.

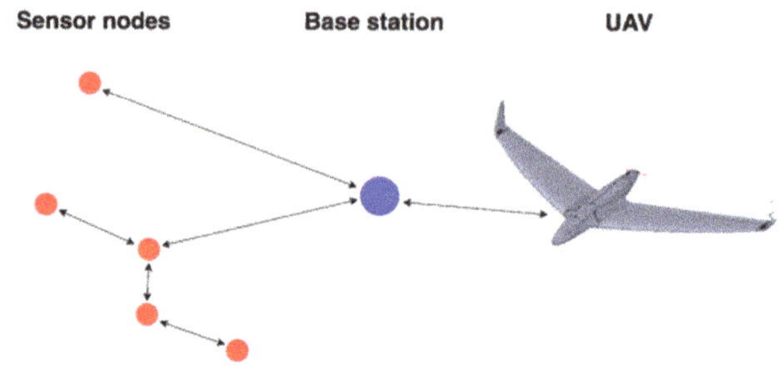

Figure 6. General configuration of UAV–WSN system implementation.

3.1. Path Tracking

We start by illustrating a nominal trajectory obtained by applying the segment tracking part of the path planning algorithm (Figure 7). The waypoints are the cluster heads (blue markers), and to each of them corresponds an update radius (solid blue line) and a communication radius (dashed black line). The first radius denotes the region in which an update of the current segment is carried out, and the second denotes the region inside which communication is possible. The starting point is chosen far away from the initial waypoint.

The algorithm provides, at each step, a heading vector which (with the use of the current position) leads to a heading angle. Together with a constant velocity value, these values are applied to a simplified 2 degrees of freedom UAV model, which is numerically integrated to provide the resultant path (solid red line). The sampling time is taken T = 1 s, and the numerical integration is done through ode45 in MATLAB 2018b.

The same scenario is carried out for the nominal case and for the case with wind disturbances (modeled by random uniform noise bounded by the interval $[-15, 15]$). The results are depicted in Figure 7, where we indeed observe a reasonable behavior of the resultant path (it passes through the waypoints neighborhoods, changes to a new segment as expected, and is smooth, at least in the nominal case).

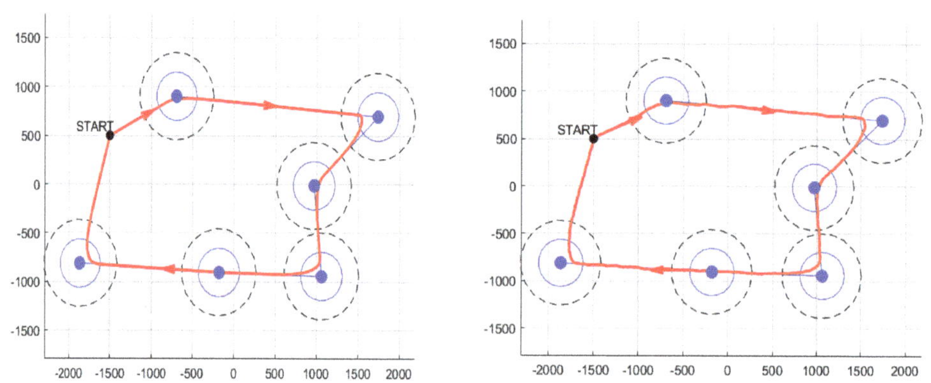

(**a**) (**b**)

Figure 7. Illustration of segment tracking: (**a**) nominal case and (**b**) with wind disturbances.

To better illustrate the scheme's performance, we show multiple runs (3 samples), each of them for various noise values. We bound the resultant paths inside a corridor of diameter d = 30 m (Figure 8).

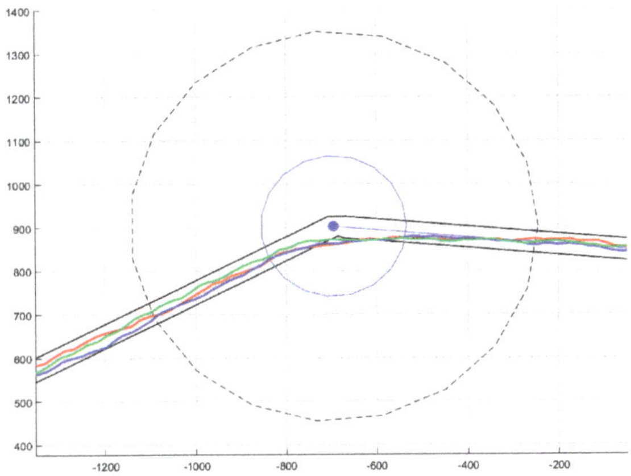

Figure 8. Illustration of trajectory tracking for multiple runs and with bounding corridor.

We observe that the resulted path does not guarantee enough time inside all communication ranges of the cluster head nodes. Specifically, we note that the 2nd and 6th waypoints (the one in the upper-most and the one in the lower-most corners) are only tangentially visited. Thus, the need for a loitering mode is clear. To better emphasize the behavior of the UAV when in loiter mode, we first show, in Figure 9, the path resulting in such a case (for both nominal and under disturbance functioning).

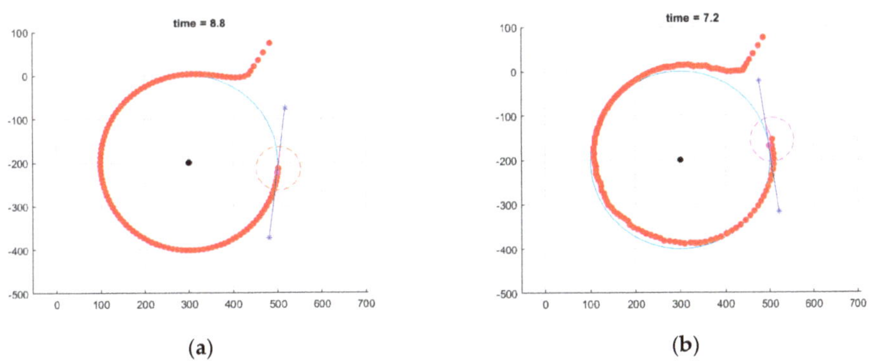

Figure 9. Illustration of loiter circle tracking: (**a**) nominal case and (**b**) with wind disturbances.

We can now integrate the full algorithm where we switch between segment and loiter modes, as needed. Specifically, in Figure 10, we consider that only waypoints 4 and 6 require the activation of the loitering mode and that the UAV stays in this mode for a fixed duration of t = 100 s. This can be obviously improved by deciding to exit the loitering mode at a later date (e.g., such that the UAV is already well-oriented towards the next waypoint).

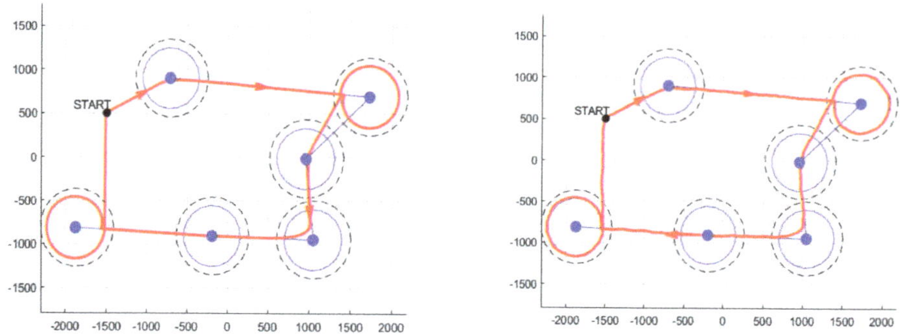

(**a**) (**b**)

Figure 10. Illustration of combined (segment and loitering circle) tracking. In all cases, the loitering circle radius was taken to be 150 m: (**a**) nominal case and (**b**) with wind disturbances.

To simulate path tracking, the NMEA (National Marine Electronics Association) Generator was used [46] (Figure 11). The path tracking, both in pattern mode (piecewise linear trajectory) and in loiter mode (circles around base stations), was simulated (Figures 12 and 13).

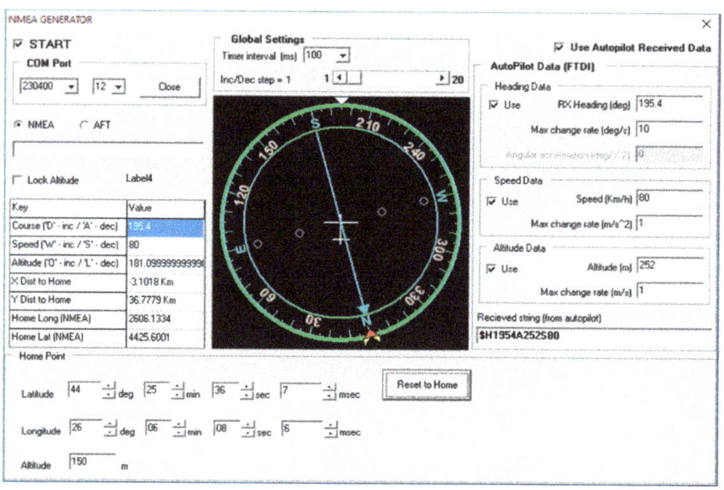

Figure 11. NMEA—based Simulator.

Figure 12. Pattern mode (tracking segments = green dashed line). Green arrow = UAV.

Figure 13. Loiter mode (tracking circles = blue). Green arrow = UAV.

3.2. Sensor Placement and Parameter Maps

UAV path planning revolves around optimizing the data collection from the cluster head with the constraint of limited mobility and hovering ability of fixed-wing type airborne platforms. To this extent, before the UAV is scheduled to visit the area, all local measurement have to be collected from the WSN at the cluster head, filtered, and aggregated, while only uploading, for example, the consensus values, confidence intervals, and outcomes of event detection and embedded alerting mechanisms.

The practical experiments at the ground sensor network level have used a sensor node deployment similar to the layout in Figure 14. In total, there are 45 nodes deployed in the field on various experimental parcels from our agronomical research institute partner. Among these nodes, six of them have the cluster head role for local collection of the sensor measurement from the neighboring nodes, as well as increased capabilities in terms of data processing, storage, and energy resources, e.g., solar panel, larger batteries, and high gain antennas for more robust operation. These are listed as blue disks in the figure, and their selection is based on the geographical coverage conditions and installation constraints.

Figure 14. Study area with the corresponding sensor nodes (red disks) and cluster heads (blue disks).

In Figure 15, a further split of the wireless sensor network is performed according to four interest zones (Zone 1–Zone 4) in the agricultural experimental area. Zone 1 contains one cluster head and 12 sensor nodes. Zone 4 contains one cluster head and six sensor nodes. For increased reliability of the data collection, in Zone 2 and Zone 3, two cluster heads are installed, with two patches of six and five sensor nodes, respectively, in the first case and two patches of six and four sensor nodes in the latter.

Based on the discussed deployment layout in the field, we present the coverage maps from the initial values for two parameters and their progression based on the implementation of the distributed agreement algorithm. In Figure 16a, the initial soil moisture values are presented. As the consensus algorithm advances in 10, 20, and 30 iterations, the coverage map is formed with increasing confidence on the joint agreement value after subsequent message exchanges. The final agreement value is stored at the cluster head to ultimately inform the decision process of the local conditions for irrigation actuation—the sensing density, in our case, is larger than the granularity of the irrigation system, which requires an average model based on the local geographical conditions.

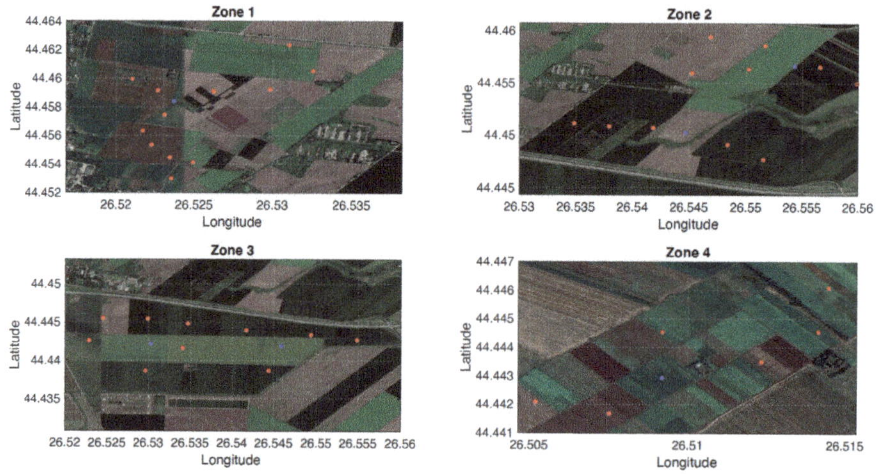

Figure 15. Location of nodes in four zones.

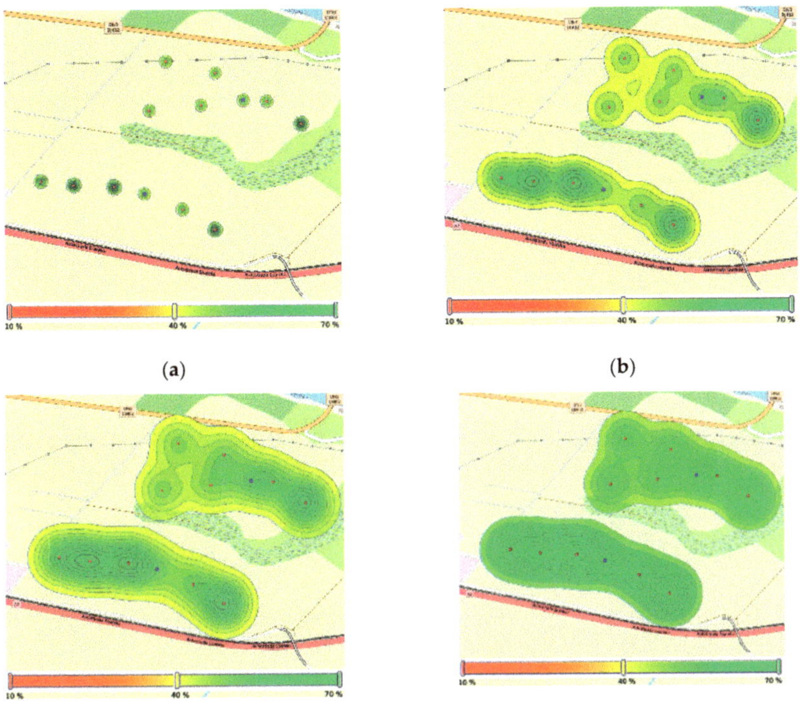

(**a**) (**b**)

(**c**) (**d**)

Figure 16. Soil moisture map in Zone 2, before and after consensus: (**a**) location of soil moisture sensors; (**b**) soil moisture map after 10 iterations; (**c**) soil moisture map after 20 iterations; and (**d**) soil moisture map after 30 iterations.

In a similar manner as for the soil moisture parameter, Figure 17 reports the initial values and the consensus progression for the air temperature parameter for Zone 2. The approach is repeated for all the parameters that can be sensed in the field. The sampling time is adapted to the process dynamics, as well as to previously reported events or external influences, e.g., weather changes, season, and expert input regarding field conditions.

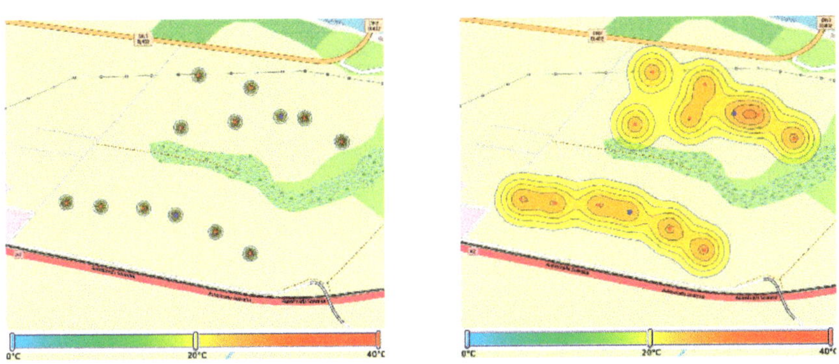

(**a**) (**b**)

Figure 17. *Cont.*

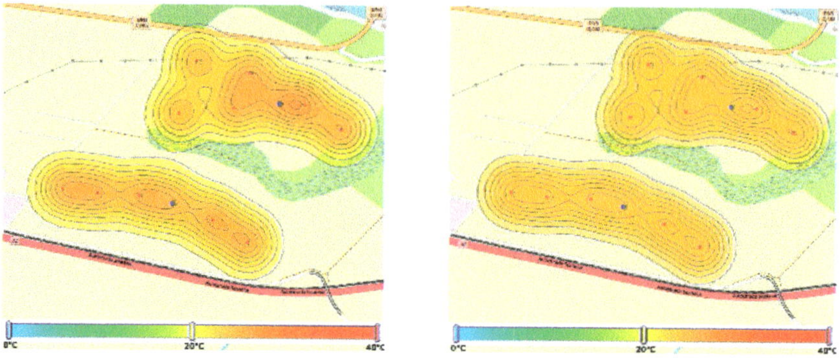

(**c**) (**d**)

Figure 17. Temperature map in Zone 2, before and after consensus: (**a**) location of temperature sensors; (**b**) temperature map after 10 iterations; (**c**) temperature map after 20 iterations; and (**d**) temperature map after 30 iterations.

3.3. Data Processing Results

As previously discussed, the primary local distributed agreement is based on consensus among the clustered sensing nodes. This allows the nodes to have a unitary representation of the measurements, under the assumption of limited variance in the geographical sensing area for one cluster. The parameters that are sampled by the nodes include: air temperature, relative humidity, soil temperature, soil moisture, and solar radiation.

Figure 18 illustrates the consensus results for two parameters: soil moisture and air temperature in a cluster of five TelosB sensor nodes. These are obtained through simulation in a Contiki/COOJA network environment starting from ground-collected values. The main insight provided by this result is in the analysis of the convergence time and convergence values in conjunction with fixed or dynamic tuning parameters. More specifically, by adjusting the communication frequency and weighting the consensus algorithm based on the sensor location and confidence levels, we can guide the algorithm with expert knowledge. This can result in acceleration of the process or in more reliable consensus values.

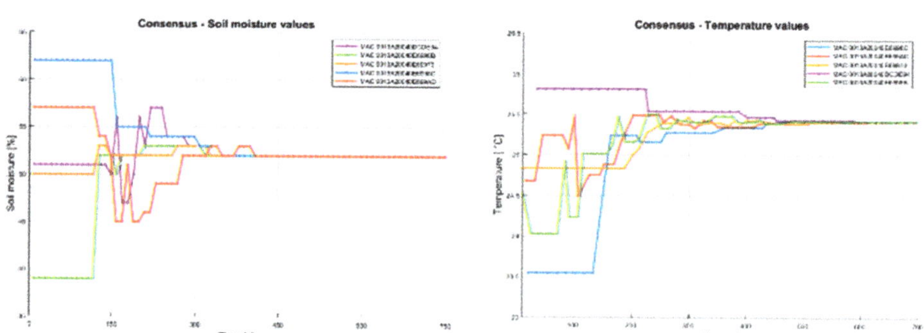

(**a**) (**b**)

Figure 18. Consensus results for: (**a**) soil moisture and (**b**) air temperature.

Once local agreement has been established, relevant data extraction is performed at the cluster head by means of the SAX method. In this case, we present the outcome for running the algorithm on a data sample of around 10 days, with the consensus values stored at 30-min intervals at one cluster head (Table 2). The variations in the SAX string length correspond to the parameterization of the method in terms of the number of segments to divide the input time series into (nseg) and the alphabet size, i.e., the discrete threshold levels numbers for classifying the processed values (alphabet_size). The number of samples of the input data is 490, for nseg = 20, corresponding to half daily patterns this is truncated to 480 as the total length of the time series must be divisible with the number of segments. Inputs are z-normalized for the computation of the assigned label. Data were collected in mid-July 2018.

Table 2. Resulting Symbolic Aggregate Approximation (SAX) strings on consensus data.

SAX Parameters	Solar Radiation	Air Temperature	Soil Temperature	Relative Humidity
nseg = 10 alphabet size = 4	bcccbccccb	bbcccbcccb	aabdccccdc	cccbbbbbbc
nseg = 10 alphabet size = 6	cdddcddddc	bcdddcdddc	aaceeddded	eddcccccce
nseg = 20 alphabet size = 4	bbbcbcbcbcbcbdbdbdab	abacbdbdadadadadbdac	aaaaaccdccccccccccdcb	dcdbdacadacadadacadc
nseg = 20 alphabet size = 6	bccdcecdbecebebebebc	bcbebfcfbebeafbfbead	aaabbdeeeededdddeeec	edebebebeaeaeaeaebed

The proposed relevant data extraction methods were evaluated from a comparative standpoint regarding the ratio between the volume of data and the data relevance. For a set of measurements, for air temperature monitoring, acquired for 10 days, 502 data points were validated and stored, totaling 2.008 kBytes. This raw data set was used for three relevant data extraction methods; the results are presented below.

Figure 19 illustrates a total of 98 relevant points extracted through the Fog computing algorithm based on change detection approach. Considering the common size of 4 bytes for floating point values, a total of approximately 400 bytes needs to be uploaded (excluding the proposed protocol frame).

For the symbolic aggregation method, two tests were performed, for two parameterizations of the SAX algorithm at opposite poles. First, Figure 19 illustrates the results for SAX algorithm adjusted for a rough representation of the time series; thus, a number of 10 characters is extracted. Considering the common size of one byte for ASCII character representation, a total of 40 bytes needs to be uploaded. Secondly, for granular SAX, Figure 19 illustrates a total of 48 points, thus totaling of 48 bytes.

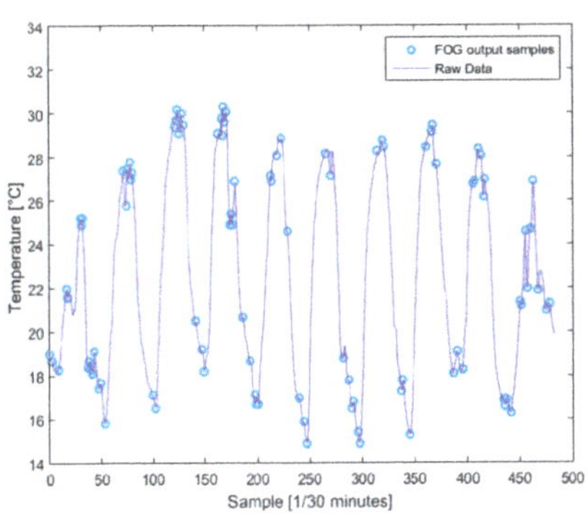

(**a**)

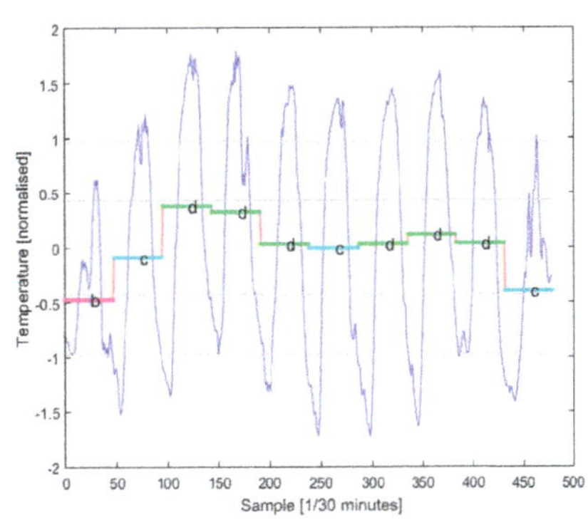

(**b**)

Figure 19. *Cont.*

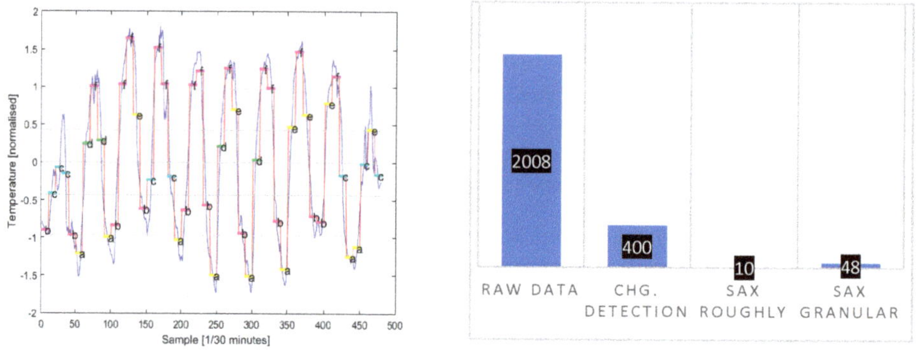

(c) (d)

Figure 19. Relevant data extraction: (**a**) change detection method; (**b**) SAX algorithm–Roughly; (**c**) SAX algorithm–Granular; and (**d**) comparative representation of data sizes achieved using the proposed relevant data extraction methods.

4. Discussion

The paper represents a significant extension of Reference [47] with further details regarding the UAV trajectory tracking and implementation of the support path planning software interfaces and illustrative path planning examples. On the data processing and deployment of the ground sensor network, the results are further elaborated upon with coverage maps, improved consensus, and relevant data extraction results. The two-stage data processing methodology presented in this paper includes a consensus algorithm for distributed agreement for sensor node patches deployed in the field alongside a relevant data extraction step based on the consensus results. The first stage is intended to ensure agreement of all the data collection entities upon the measured parameters, as well as to increase data quality by limiting the effect of sending upstream erroneous sensor readings. The second stage aims to optimize the data collection time at the interface between the cluster head and the UAV acting as a data mule. Based on the compressed representation of SAX segments, the results can be expanded and further processed at the decision level, in the cloud. At the higher abstract layer in the cloud, the results presented in Table 2 can be interpreted using state-of-the-art text analytics tools. This is useful for quantitative assessment of univariate sequences, as well as correlations between multivariate string series. The character frequencies and recurring subsequences for certain parameters might be indicators for evolving phenomena at the ground level.

Potential drawbacks of the integrated system are related to the increased complexity for multi-level data processing, communication, and interoperability constraints between the aerial platform and the ground sensors. Increased administrative requirements have to be complied with, e.g., approving flight plans for each UAV mission, along with maintenance requirements that can stem from outdoor deployment of the nodes. We consider, however, that the benefits outweigh the discussed drawbacks of such a system.

5. Conclusions

The paper illustrated a case study for collaborative UAV–WSN operation in large scale monitoring for precision agriculture. The algorithms, techniques, and tools to enable seamless interoperability between the two domains are illustrated. Key contributions are argued in the design of optimized trajectories for UAV-enabled field data collection and for in-network data processing that allows efficient use of limited ground sensor network resources. Particularly, we propose combined segment and loiter tracking modes which balance between path length and time spent in the neighborhood of a cluster head. By passing the raw sensor readings through multiple hierarchical data processing steps, the quality of the extracted information is increased, as well as its timeliness, given the fact that reduced communication burden allows lower network-wide latency for decision-making. The role of the UAV platform is critical to support large scale monitoring and data collection applications in precision agriculture as it reduces the reliance of third-party communication and computing infrastructure that might not be readily available in the field or pose increased costs.

Extensive field evaluation is planned for validation of the impact of such a system for crop management. The main challenges for such a collaborative system are the following: sensing covering, communication covering by the hybrid UAV–ground WSN system, energy efficiency, and computing efficiency.

Author Contributions: D.P. conceived the paper and revealed the collaborative function. F.S. contributed to UAV path analysis, G.S. studied the data acquisition and processing functions. C.D. studied the system applications. L.I. selected the references, state of the art and edited the paper. All authors have read and agreed to the published version of the manuscript.

Funding: This research awas funding by University POLITEHNICA of Bucharest.

Acknowledgments: The work was supported by the project MUWI (Integrated intelligent monitoring system, UAV–WSN–IoT, for precision agriculture), 1224/2018, subsidiary NETIO.

Conflicts of Interest: The authors declare no conflict of interest.

References

1. Wolfert, S.; Ge, L.; Verdouw, C.; Bogaardt, M.-J. Big Data in smart farming—A review. *Agric. Syst.* **2017**, *153*, 69–80. [CrossRef]
2. Piotrowski, K.; Sojka-Piotrowska, A.; Stamenkovic, Z.; Kraemer, R. IHPNode platform as a base for precision farming and remote diagnosis in agriculture. In Proceedings of the 24th Telecommunications forum TELFOR, Serbia, Belgrade, 22–23 November 2016; pp. 1–5.
3. Valente, J.; Sanz, D.; Barrientos, A.; Cerro, J.D.; Ribeiro, A.; Rossi, C. An air-ground wireless sensor network for crop monitoring. *Sensors* **2011**, *11*, 6088–6108. [CrossRef]
4. Di Gennaro, S.F.; Matesea, A.; Gioli, B.; Toscano, P.; Zaldei, A.; Palliotti, A.; Genesio, L. Multisensor approach to assess vineyard thermal dynamics combining high resolution Unmanned Aerial Vehicle (UAV) remote sensing and wireless sensor network (WSN) proximal sensing. *Sci. Hortic.* **2017**, *221*, 83–87. [CrossRef]
5. Polo, J.; Hornero, G.; Duijneveld, C.; García, A.; Casas, O. Design of a low-cost Wireless Sensor Network with UAV mobile node for agricultural applications. *Comput. Electron. Agric.* **2015**, *119*, 19–32. [CrossRef]
6. Primicerio, J.; Matese, A.; Gennaro, S.D.; Albanese, L.; Guidoni, S.; Gay, P. Development of an integrated, low-cost and opensource system for precision viticulture: From UAV to WSN. In Proceedings of the EFITA-WCCA-CIGR Conference Sustainable Agriculture through ICT Innovation, Torino, Italy, 23–27 June 2013; pp. 1–6.
7. Drones—Regulatory Framework Background/EASA. Available online: http://www.easa.europa.eu (accessed on 1 December 2019).
8. FAA Reauthorization Act of 2018/Subtitle B—Unmanned Aircraft Systems. Available online: https://www.faa.gov (accessed on 1 December 2019).
9. Barbedo, J.G.A. A review on the use of Unmanned Aerial Vehicles and imaging sensors for monitoring and assessing plant stresses. *Drones* **2019**, *3*, 40. [CrossRef]
10. Shafi, U.; Mumtaz, R.; García-Nieto, J.; Hassan, S.A.; Zaidi, S.A.R.; Iqbal, N. Precision agriculture techniques and practices: From considerations to applications. *Sensors* **2019**, *19*, 3796. [CrossRef]
11. Popescu, D.; Ichim, L.; Stoican, F. Unmanned Aerial Vehicle Systems for remote estimation of flooded areas based on complex image processing. *Sensors* **2017**, *17*, 446. [CrossRef]
12. Xu, Y.; Xiao, L.; Yang, D.; Cuthbert, L.; Wang, Y. Energy-efficient UAV communication with multiple GTs based on trajectory optimization. *Mob. Inf. Syst.* **2018**, *2018*, 5629573. [CrossRef]
13. Sylvester, G. *E-Agriculture in Action: Drones for Agriculture*; Food and Agriculture Organization of the United Nations and International Telecommunication Union: Bangkok, Thailand, 2018; Available online: http://www.fao.org/documents/card/en/c/I8494EN/ (accessed on 1 December 2019).

14. Matese, A.; Di Gennaro, S.F. Practical applications of a multisensor UAV platform based on multispectral, thermal and RGB high resolution images in precision viticulture. *Agriculture* **2018**, *8*, 116. [CrossRef]
15. Ballesteros, R.; Ortega, J.F.; Hernandez, D.; Moreno, M.A. Onion biomass monitoring using UAV-based RGB imaging. *Precis. Agric.* **2018**, *19*, 840–857. [CrossRef]
16. Karydas, C.; Gewehr, S.; Iatrou, M.; Iatrou, G.; Mourelatos, S. Olive plantation mapping on a sub-tree scale with object-based image analysis of multispectral UAV data; Operational potential in tree stress monitoring. *J. Imaging* **2017**, *3*, 57. [CrossRef]
17. Johansen, K.; Raharjo, T.; McCabe, M.F. Using multi-spectral UAV imagery to extract tree crop structural properties and assess pruning effects. *Remote Sens.* **2018**, *10*, 854. [CrossRef]
18. Gómez-Candón, D.; Torres-Sanchez, J.; Labbé, S.; Jolivot, A.; Martinez, S.; Regnard, J. Water stress assessment at tree scale: High-resolution thermal UAV imagery acquisition and processing. *Acta Hortic.* **2017**, *1150*, 159–166. [CrossRef]
19. Ribeiro-Gomes, K.; Hernandez-Lopez, D.; Ortega, J.F.; Ballesteros, R.; Poblete, T.; Moreno, M.A. Uncooled thermal camera calibration and optimization of the photogrammetry process for UAV applications in agriculture. *Sensors* **2017**, *17*, 2173. [CrossRef] [PubMed]
20. Tetila, E.C.; Machado, B.B.; Belete, N.A.S.; Guimarães, D.A.; Pistori, H. Identification of Soybean foliar diseases using Unmanned Aerial Vehicle images. *IEEE Geosci. Remote Sens. Lett.* **2017**, *14*, 2190–2194. [CrossRef]
21. Stamenković, Z.; Randjić, S.; Santamaria, I.; Pešović, U.; Panić, G.; Tanasković, S. Advanced Wireless Sensor Nodes and networks for agricultural applications. In Proceedings of the 24th Telecommunications Forum (TELFOR), Belgrade, Serbia, 22–23 November 2016; pp. 1–8.
22. Popescu, D.; Stoican, F.; Stamatescu, G.; Chenaru, O.; Ichim, L. A survey of collaborative UAV-WSN systems for efficient monitoring. *Sensors* **2019**, *19*, 4690. [CrossRef]
23. Moreno-Moreno, C.D.; Brox-Jiménez, M.; Gersnoviez-Milla, A.A.; Márquez-Moyano, M.; Ortiz-López, M.A.; Quiles-Latorre, F.J. Wireless Sensor Network for sustainable agriculture. *Proceedings* **2018**, *2*, 1302. [CrossRef]
24. Husemann, C.; Novković, N. Farm management information systems: A case study on a german multifunctional farm. *Econ. Agric.* **2014**, *2*, 1–13. [CrossRef]
25. Köksal, Ö.; Tekinerdogan, B. Architecture design approach for IoT-based farm management information systems. *Precis. Agric.* **2019**, *20*, 926–958. [CrossRef]
26. Kumar, A.; Ilango, P. The impact of Wireless Sensor Network in the field of precision agriculture: A review. *Wirel. Pers. Commun.* **2017**, *98*, 685–698. [CrossRef]
27. Keshtgary, M.; Deljoo, A. An efficient wireless sensor network for precision agriculture. *Can. J. Multimed. Wirel. Netw.* **2012**, *3*, 1–6.
28. Nikolidakis, S.A.; Kandris, D.; Vergados, D.D.; Douligeris, C. Energy efficient automated control of irrigation in agriculture by using wireless sensor networks. *Comput. Electron. Agric.* **2015**, *113*, 154–163. [CrossRef]
29. Yousif, M.E.R.; Ghafar, K.; Zahari, R.; Lim, T.H. A rule-based smart automated fertilization and irrigation systems. In Proceedings of the Ninth International Conference on Graphic and Image Processing (ICGIP 2017), Qingdao, China, 14–16 October 2017.
30. Popescu, D.; Dragana, C.; Stoican, F.; Ichim, L.; Stamatescu, G. A collaborative UAV–WSN network for monitoring large areas. *Sensors* **2018**, *18*, 4202. [CrossRef] [PubMed]
31. Vasisht, D.; Kapetanovic, Z.; Won, J.; Jin, X.; Chandra, R.; Kapoor, A.; Sinha, S.N.; Sudarshan, M.; Stratman, S. Farmbeats: An IoT platform for data-driven agriculture. In Proceedings of the 14th USENIX Conference on Networked Systems Design and Implementation (NSDI), Boston, MA, USA, 27–29 March 2017; pp. 515–528.
32. Lysenko, V.; Opryshko, O.; Komarchuk, D.; Pasichnyk, N.; Zaets, N.; Dudnyk, A. Information support of the remote nitrogen monitoring system in agricultural crops. *Int. J. Comput.* **2018**, *17*, 47–54.
33. Stamatescu, G.; Stamatescu, I.; Drăgana, C.; Popescu, D. Large scale heterogeneous monitoring system with decentralized sensor fusion. In Proceedings of the IEEE 8th International Conf. on Intelligent Data Acquisition and Advanced Computing Systems (IDAACS), Warsaw, Poland, 24–26 September 2015; pp. 2–5.
34. Dragana, C.; Stamatescu, G.; Mihai, V.; Popescu, D. Evaluation of cluster formation algorithm in large scale wireless sensor network. In Proceedings of the 9th IEEE Intl Conf on Intelligent Data Acquisition and Advanced Computing Systems (IDAACS), Bucharest, Romania, 21–23 September 2017; pp. 859–863.
35. Anawar, M.R.; Wang, S.; Azam Zia, M.; Jadoon, A.K.; Akram, U.; Raza, S. Fog computing: An overview of big IoT data analytics. *Wirel. Commun. Mob. Comput.* **2018**, *2018*, 1–22. [CrossRef]

36. Yang, Q.; Yoo, S. Optimal UAV path planning: Sensing data acquisition over IoT sensor networks using multi-objective bio-inspired algorithms. *IEEE Access* **2018**, *6*, 13671–13684. [CrossRef]
37. Goudarzi, S.; Kama, N.; Anisi, M.H.; Zeadally, S.; Mumtaz, S. Data collection using Unmanned Aerial Vehicles for Internet of Things platforms. *Comput. Electr. Eng.* **2019**, *75*, 1–15. [CrossRef]
38. Zhan, C.; Zeng, Y.; Zhang, R. Energy-efficient data collection in UAV enabled Wireless Sensor Network. *IEEE Wirel. Commun. Lett.* **2018**, *7*, 328–331. [CrossRef]
39. Stoican, F.; Prodan, I.; Popescu, D. Flat trajectory generation for way-points relaxations and obstacle avoidance. In Proceedings of the 23rd Mediterranean Conference on Control and Automation (MED), Torremolinos, Spain, 16–19 June 2015; pp. 695–700.
40. Sager, S. Numerical Methods for Mixed-Integer Optimal Control Problems. Ph.D. Thesis, Universitat Heidelberg, Der andere Verlag Tonning, Germany, 2005.
41. Liu, B.; Zhu, H. Energy-effective data gathering for UAV-aided Wireless Sensor Networks. *Sensors* **2019**, *19*, 2506. [CrossRef]
42. Santamaria, A.F.; Raimondo, P.; Tropea, M.; De Rango, F.; Aiello, C. An IoT surveillance system based on a decentralised architecture. *Sensors* **2019**, *19*, 1469. [CrossRef]
43. Alighanbari, M.; Kuwata, Y.; How, J. Coordination and control of multiple UAVs with timing constraints and loitering. In Proceedings of the American Control Conference, Denver, CO, USA, 4–6 June 2003; pp. 5311–5316.
44. Blanchini, F.; Miani, S. *Set-Theoretic Methods in Control*; Springer: Cham, Switzerland, 2008; ISBN 978-0-8176-3255-7.
45. Lin, J.; Keogh, E.; Wei, L.; Lonardi, S. Experiencing SAX: A novel symbolic representation of time series. *Data Min. Knowl. Discov.* **2007**, *15*, 107–144. [CrossRef]
46. NMEA Generator. Available online: https://nmeagen.org (accessed on 1 September 2019).
47. Popescu, D.; Stoican, F.; Ichim, L.; Stamatescu, G.; Dragana, C. Collaborative UAV-WSN system for data acquisition and processing in agriculture. In Proceedings of the 10th IEEE International Conference on Intelligent Data Acquisition and Advanced Computing Systems: Technology and Applications (IDAACS), Metz, France, 18–21 September 2019; pp. 519–524.

sensors

MDPI

Article

EUROPA: A Case Study for Teaching Sensors, Data Acquisition and Robotics via a ROS-Based Educational Robot

Georgios Karalekas, Stavros Vologiannidis * and John Kalomiros

Department of Computer, Informatics and Telecommunications Engineering, International Hellenic University, 62124 Serres, Greece; karalekasgeorge@gmail.com (G.K.); ikalom@ihu.gr (J.K.)
* Correspondence: svol@ihu.gr

Received: 27 March 2020; Accepted: 23 April 2020; Published: 26 April 2020

check for updates

Abstract: Robots have become a popular educational tool in secondary education, introducing scientific, technological, engineering and mathematical concepts to students all around the globe. In this paper EUROPA, an extensible, open software and open hardware robotic platform is presented focusing on teaching physics, sensors, data acquisition and robotics. EUROPA's software infrastructure is based on Robot Operating System (ROS). It includes easy to use interfaces for robot control and interaction with users and thus can easily be incorporated in Science, Technology, Engineering and Mathematics (STEM) and robotics classes. EUROPA was designed taking into account current trends in educational robotics. An overview of widespread robotic platforms is presented, documenting several critical parameters of interest such as their architecture, sensors, actuators and controllers, their approximate cost, etc. Finally, an introductory STEM curriculum developed for EUROPA and applied in a class of high school students is presented.

Keywords: educational robotics; data acquisition; sensors; ROS; STEM

---

1. Introduction

Robotics represents an innovative field that encompasses various scientific domains, from physics and electronics to mechanical engineering, mathematics and computer programming. The vast field of artificial intelligence is also relevant to autonomous robots. Educational robotics is a rapidly evolving multidisciplinary domain that brings together educators, companies and researchers in an effort to create a new learning environment in schools and universities. Rooted in Papert's seminal ideas on computational thinking using LOGO programming and Mindstorms [1], educational robotics is increasingly becoming popular in the classroom. It is supportive in teaching Science, Technology, Engineering and Mathematics (STEM) [2,3] and in some cases it transcends the traditional science border and becomes supportive of artistic activities (STEAM) [4,5].

Introducing robots in the classroom can become a suitable tool to instill new skills in young learners and students; besides teamwork and creativity, students can develop interest in practical concepts in physics and mathematics and get acquainted with topics in engineering [6]. Robotics can help teachers present the concept of system engineering and control. In addition, it can motivate young students towards STEM-oriented career paths, which has become important following the expansion of technology markets and their demand for engineering graduates. Interacting with robots can even be beneficial for children in a kindergarten [7,8] and it can play an important therapeutic role in special education [9].

Innovative learning based on robotics also brings about the need to develop new curricula for schools and universities, to cover gaps in documentation for teachers and students and to develop new products in the form of simple, low-cost mobile platforms, according to the educational level of the target group. On the other hand, introducing robotics courses in schools calls for the parallel development of methods for the assessment of new educational technologies and the dissemination of their results [10]. As robot-based technologies are becoming mainstream in schools and universities, educational robotics is gaining its own spin and status among researchers, markets and educators, with new emerging conferences [11], special issues [12] and products [13,14] and with a boost in the relevant literature [15].

Within the constructivist approach of Piaget [16] and Papert [17], constructing the robot can be considered an integral part of the learning procedure. Construction not only stimulates the creativity and enthusiasm of young learners through an open-ended, problem-solving process in the real world, but it also instills technological literacy and better understanding of the different parts that make up a robot as an engineering system. This is especially true for primary and secondary (K-12) education [18]; however, it can also find application in college or university education, where lab exercises on robotics often include a basic assembly of a simple robot, like a mobile cart driven differentially. Several of the educational bots currently available as market products allow some level of assembly of the robot from parts, while others encourage extensions of a ready product.

The main challenges when designing a new robotic platform for education are component accessibility, flexibility and cost. It is preferable to design platforms based on commodity components that can be easily accessed in the market and replaced when needed. The platform should be flexible enough to adjust to different teaching scenarios. In part, this means that an educational robot should best follow a modular architecture in terms of sensors and accessories and in terms of software, especially in order to span different curricula. Finally, a low-cost platform makes an investment in robotic technology more plausible for a large classroom, where each small group of three or four students should share a robot with its accessories and build several projects around it.

Using open-source software and open hardware in designing an educational system is important, especially for high school secondary education and for university courses. Open hardware, like Arduino Uno [19], with its free programming environment and community support [20] can increase the level of student creativity and engagement in a robotics project. Similarly, the Raspberry Pi [21], although it does not exactly represent open hardware, is supported by a large community, runs a version of the Linux Operating System and can be programmed using Python. Python is widely taught in Informatics lessons in various high school curricula, as is the case in Greece. Hardware boards like the above provide user-friendly input/output support and can be easily adopted for other technology-oriented extra-curricular activities, beside educational robotics.

The power of open software in educational robotics is best exemplified by the Robot Operating System (ROS). ROS [22] is a middleware that runs on Linux and recently on Windows 10 and has become a standard for robotics, in industry, education and research. It provides easy access to complex software components and communicates with a great variety of hardware, like sensors and actuators. It allows the robot integration with tools for simulation and visualization [23–25] and with libraries for robotic vision, artificial intelligence and Simultaneous Localization and Mapping (SLAM). These powerful functionalities transform the robot from a simple programmable automatic system to a true autonomous intelligent device, compliant with the technology of the Internet of Things (IoT).

A number of educational platforms have been presented as market products and have been introduced in various levels of education. From Beebot [26,27] to Thymio II [28,29] to Scribbler 3 [30] and LEGO EV3 [31–33], the educational market has provided teachers with ingenious tools to devise innovative lessons on almost everything. Activities range from exhibiting a practical algorithm in kindergarten to teaching concepts on motion and automation to understanding basic programming and the role of sensors and actuators in a control loop. More advanced platforms, like the epuck [33,34], the Turtlebot [35] and the Duckietown [36], introduce students to the use of single board computers, path planning and environmental mapping. They use cameras and artificial intelligence for object recognition and are suitable for research on advanced stochastic algorithms for localization and mapping. More industrial-like robotic platforms, like DaNI and VEGA, are often adopted for the needs of the postgraduate level and for research [37]. In the same category, the Pioneer mobile platforms by Adept have been very popular for autonomous navigation research but they are gradually replaced by a line of ROS based autonomous mobile platforms, like the Leo Rover [38]. Finally, pure industrial grade robot platforms, like the Robotnik^{TM} Summit-XL [39] or the Husky and Jackal unmanned mobile bases by Clearpath Robotics^{TM} [40], are fully ROS based customizable platforms, suitable for research projects and industrial or agricultural applications.

The contribution of this paper is twofold: first, we present a comprehensive review of the state of the art on educational robotic platforms through K12 to college and university and second, we present EUROPA, a new educational mobile platform based on ROS, which has been developed following the main guidelines stipulated above: constructivist approach, accessibility of parts, modular flexibility and open hardware and software technology. The platform has been introduced in a secondary school class following the Greek educational system and has been positively assessed by students and tutors. A short curriculum is also proposed for the blending of robotic technology with STEM teaching, in secondary school.

The rest of the paper is structured as follows. In Section 2 we present a comprehensive state-of-the-art review on the technology of educational mobile platforms through various levels of education. In Section 3 the hardware and software architecture of the proposed EUROPA platform with its ROS software architecture is presented. In Section 4 EUROPA is studied as a paradigm of introducing a robot in class and the assessed curriculum is outlined. A configuration of EUROPA for teaching more advanced robotics is also proposed, aiming to lessons on autonomous driving, typically applying to university education. Finally, Section 5 concludes the paper.

2. State-of-the-Art Educational Platforms

In this section, several well-known mobile platforms used in education are presented, starting with bots that have been adopted to teach computational thinking and basic notions of programming in elementary school, proceeding with platforms that can be used in STEM classes which enhance engineering literacy in high school and ending with projects designed to teach autonomy or test computer vision and navigation algorithms in university and research. Of course, this presentation cannot be exhaustive, since there is a large number of products, some very successful, others very promising, several of low cost and some based on open hardware/software. However, we took care to include those platforms that appear often in the literature on educational robotic technology or are promising in our opinion to lead a trend in a specific educational level. A comparative examination of the technology and specifications of such successful platforms can indicate how the next generation of educational robotic technology is going to evolve. A reference to most of the products that are not directly presented in this section can be found in the proposed literature. Humanoids and torsos, like NAO [41], Pepper [42] or the Robotis OP3 [43] are becoming part of the educational robotics ecosystem; however, this review is limited to wheeled mobile platforms with a relatively low degree of complexity and with affordable cost in the context of school/college education.

Table 1 lists fourteen widely used educational platforms as well as our proposed EUROPA robot. The table presents the basic technologies supported by each platform and the level of education they best fit in. The current approximate cost of the platform is also given, as it is suggested by the distributor. In the last column, a reference to the literature presenting the platform capabilities or its exemplary use in class is provided. Each one of the listed systems is illustrated in Figure 1. One industrial platform, the Summit-XL, is also presented as a comparative reference.

The Beebot represents a category of toy robots appropriate for teaching introductory notions of control. It illustrates directional language and following steps in problem solving, like in a maze. It is used widely in kindergarten and elementary education with exciting results [26]. Being a toy rather than a well-defined robot vehicle, it is not well documented with regard to its mechanical and electronic specifications.

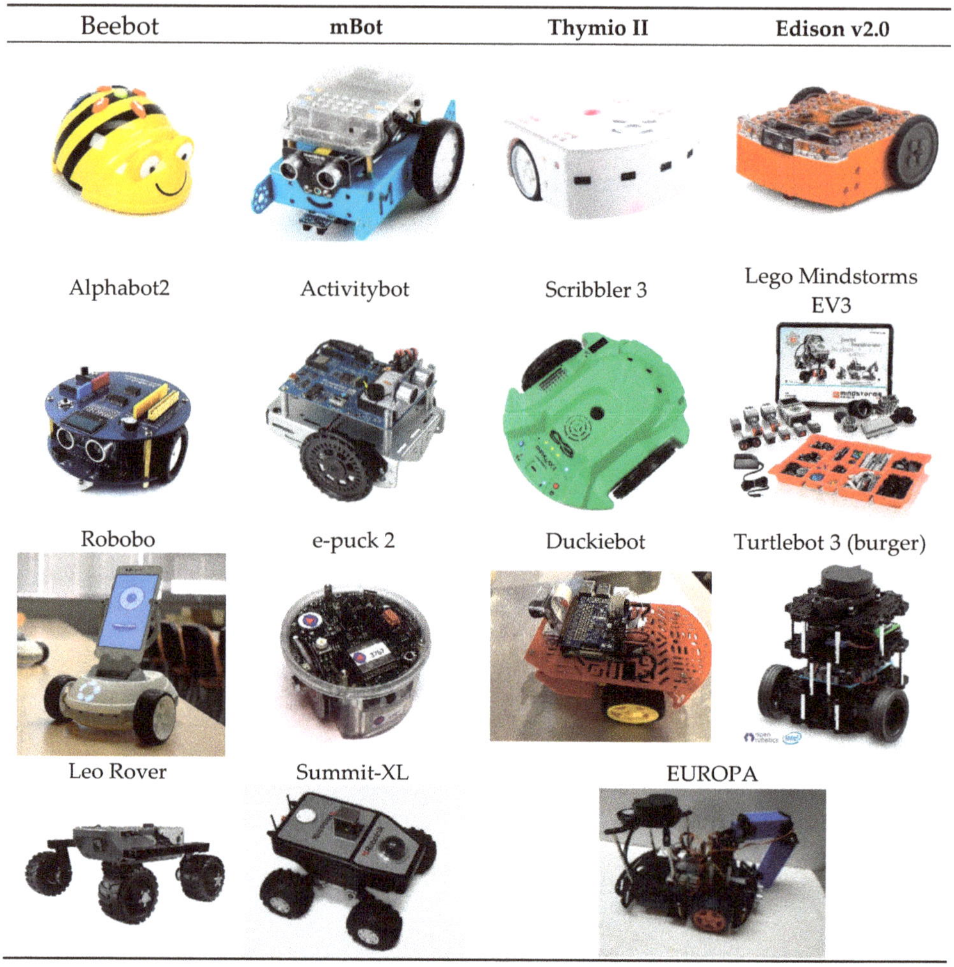

Figure 1. Images of the mobile robot platforms listed in Table 1.

The mBot is an introductory small robot by Makeblock [44]. It is based on a light metal chassis and can be assembled from parts. It can be programmed either by a block-based graphical programming interface based on scratch or using the Arduino Integrated Development Environment (IDE). Makeblock provides STEM teaching case studies in its webpage.

Thymio II is a versatile open platform suitable for all levels of K12 education, best documented with activities for elementary school. It supports six basic "behaviors", allowing obstacle avoidance, line following, hand following, etc [28]. It is based on a PIC24 microcontroller unit with an H-bridge for motor driver. It features a number of sensors, like accelerometer, thermometer and infrared proximity sensors for obstacle avoidance. Its basic actuators are two basic motors driven differentially, a loud-speaker and leds. The platform is expandable using accessories and is poised to evolve into a STEM teaching tool for higher grades or possibly into a ROS platform [45].

Table 1. Most widespread mobile educational platforms: target group, cost and technology.

Bot Name	Level of Education	Appro-ximate Cost (€)	Open Source HW/SW	Sensors	Actuators/ Kinematic Model	ROS	Controller/CPU	Programming Tools	Data Communication	Ref
Bee-bot /Colby	Kindergarten/Elen	65	NO/NO	Bs	?/DD	NO	?	Buttons	NO	[26,27]
mBot	Elementary/ secondary	120	NO/YES	B, UDS, LF, LS, IR	DC gear motor plastic, RGB LEDs, buzzer/DD	NO	ATmega328/ motor driver	Block-based/ Arduino IDE	BT	[44]
Thymio II	Elementary/ secondary	170	YES/YES	Bs, IR, LF, Th, Acc, Mic	DC gear motor plastic, LEDs, Loud speaker/ DD	NO	PIC24FJ128GB/ motor driver	Block/Visual /text based (Aseba)	2.4 GHz, protocol 802.15.4	[28,29]
Edison	Elementary/ secondary	60	NO/NO	Bs, IR, LS, LF, Mic, OE	DC gear motor, LEDs, buzzer/DD	NO	Freescale 8-bit MC9S08PA16	Block-based/ Scratch/EdPy	IR	[46]
Scribbler 3	Elementary/ secondary	200	NO/NO	LS, LF, IR, OE	DC gear motor, LEDs/DD	NO	Propeller P8X32A	Block-based programming	USB	[30]
LEGO EV3	secondary	500	NO/NO	TS, CS, GS, UDS	Compact Gear motors/DD	NO	ARM9	EV3 icon-based software	BT, Wi-Fi	[32]
AlphaBot2	secondary	90–125	YES/YES	UDS, IR, LF, Camera*	N20 micro gear motor, RGB LEDs/DD	NO	Arduino or Raspberry Pi zero or BBC micro:bit	Arduino IDE or Python	BT, IR	[47]
ActivityBot	secondary	200	NO/NO	UDS, LS, TS, IR, OE	High speed 360° servos/DD	NO	Propeller P8X32A	Block-based graphical/C	USB	[30]
Epuck 2	Secondary/ higher	1200	YES/YES	IR, acc, gyro, mic, camera, ToF	Stepper motors, LEDs, Loud-seaker/DD	YES	STM32F4 ARM Cortex M4	Free C compiler	BT	[33,34]
Robobo	Secondary/ higher	450	NO/YES	Camera, acc, gyro, GPS, magn, IR, LS	DC gear motor, LEDs/DD	YES	Smartphone + PIC32 (low-level control)	Scratch/ Python/ ROS	Wi-Fi	[48–50]
EUROPA II	Secondary/ higher	120 or 300**	YES/YES	UDS, IR, OE, Camera, LIDAR**	DC gear motor, robotic arm, LEDs/DD	YES	Raspberry Pi	Python/ROS/ OpenCV	Wi-Fi	[51]
Turtlebot 3 (burger)	higher	800	YES/YES	Camera, LIDAR, acc, gyro, magn	DYNAMIXEL AX gear motor + driver/DD	YES	Raspberry Pi + OpenCR	Block-based/ ROS/Python	Wi-Fi	[35,52]
Duckiebot	higher	150	YES/YES	Fish-eye camera for Raspberry Pi	DC gear motor/DD	YES	Raspberry Pi 2	ROS programming in C/Python	Wi-Fi	[13,36]
Leo Rover	Research/ industry	2500	YES/YES	Fish-eye camera, wheel encoders	4x DC gear motor	YES	Raspberry Pi + Core2 ROS (low level)	ROS Programming	Wi-Fi	[38]
Summit-XL	Industry	15000	YES/YES	3D camera, Laser scanner + optional sensors	4x DC gear motor/skid steering	YES	Intel processor/PC	ROS Programming	Wi-Fi	[39]

?: Not documented, DD: Differential Drive, B (s): Button (s), UDS: Ultrasonic Distance Sensor, LF: Line Following sensor, LS: Light Sensor, IR: Infrared Proximity Sensor, TS: Touch sensor, CS: Color sensor, Acc: accelerometer, Mic: Microphone, OE: Optical Encoder, ToF: Time of Flight distance sensor, magn: magnetometer, BT: Bluetooth *: in RPi version, **: University Edition.

Edison is primarily a very affordable mobile platform for teaching STEM [46]. It is equipped with a similar range of sensors and actuators, like Thymio, although it is not as "moody" and easy to personalize as Thymio and it does not belong to the open hardware and software camp. It can avoid obstacles and track a line using IR sensors and can respond to sound or play music using the integrated sound/buzzer module. It can interact with other robots using light signals. It can be programmed using three different versions of a programming environment: EdBlocks for programming with icons, EdScratch, using a block based visual programming style and the text based EdPy, which is a version of the Python language.

Lego Mindstorms EV3 [32] is a kit for educational robotics, consisting of a programmable brick and a set of motors, sensors and TECHNIC elements that can be used to assemble the robot. EV3 continues the line of Mindstorms NXT, featuring a more powerful ARM9 processor and 64MB RAM. It supports Wi-Fi and Bluetooth connectivity and can be programmed using the custom programming environment Lego Mindstorms EV3 Home Edition, which is based on a block-based graphic language originating from LabVIEW, by National Instruments. This platform is widely used in competitions.

Alphabot2 is a small mobile platform by Waveshare [47] that comes in various flavors. In its cost-effective version it hosts an Arduino controller, while it can also come with a Raspberry Pi or with a BBC micro:bit microcontroller. An ultrasonic distance sensor is used in all variations for obstacle avoidance. Alphabot2 represents open hardware and can be programmed using the Arduino IDE or Python scripts, depending on the controller.

Scribbler 3 and Activitybot are robots powered by the well-known Propeller CPU made by Parallax [30]. Scribbler 3 is a robust plastic platform suitable mostly for elementary education, which can be programmed using a block-based programming language. Activitybot features a metallic chassis and a small breadboard for adding sensors and other circuitry. Besides the block-based graphical environment, Activitybot can also be programmed in C.

The e-puck 2 [33,34] is a small differential wheeled robot designed for research and education. It is powered by a STM32F4 microcontroller and features many sensors, like IR and Time of Flight distance sensor, IMU, color sensor, etc. It is also suitable to study swarm and evolutionary robotics. It supports C programming and ROS libraries.

The Robobo [48,49] is different from the above and represents an interesting experiment by the University of Coruña. It consists of a mobile base and an attached smartphone. It makes use of the CPU power of the smartphone and of sensors incorporated in it, mainly cameras, gyroscope, accelerometer and GPS. The robot can be programmed using a Scratch web-based editor or a text-based language and aims to introduce lessons on autonomy to secondary school students [50].

The Turtlebot 3 [35] is a relatively low-priced, small size differentially driven mobile platform based on ROS. It is an open source collaboration project by several partners [14] and it is assembled from high quality modular parts. It is based on 3D-printed expandable chassis and is controlled by an effective controller and Single Board Computer. The main sensor of the Turtlebot is a low-cost LIDAR that is able to perform navigation tasks and SLAM. It can also be expanded by other sensors, like RGB and RGBD camera, supported by ROS software modules. It can be used as a mobile manipulator, by attaching a manipulator module. The Turtlebot has been used successfully in graduate education and research [52].

The Duckietown [36] is an open project proposed by a MIT team, intended for teaching robot autonomy or individual aspects of autonomous driving, like vision or nonlinear control, at a graduate or postgraduate level. It consists of the Duckiebots, which are open inexpensive differentially driven mobile bots and a model environment representing a miniature town with roads, signs and inhabitants, assembled from modular tiles. The sole sensor of the Duckiebot is a monocular camera. Vision based algorithms are responsible for lane detection, sign or object recognition and localization of the robot in the Duckietown [13]. More advanced algorithms allow path planning using metric and topological maps as well as vision-based Simultaneous Localization and Mapping. The system supports ROS for data transfer between software nodes. It can be expanded for the study of multirobot behavior. The cost given in Table 1 refers to a single bot without the Duckietown.

The Leo Rover [38] is a robust open source platform designed for autonomy research in outdoor environment. It is customizable by add-ons, like a manipulator, GPS module, camera, IMU, etc. The robot is driven by four independent DC gear motors with suspension system and it is powered by Raspberry Pi and a Core 2 ROS driver board. Although it represents an open platform with a GitHub repository, it requires extensive programming by the developer for the execution of every specific task. Therefore, its scope is different than that of educational boards.

Finally, the Summit-XL platform by Robotnik [39] is a versatile strong frame, based on a four-wheel skid-steering configuration, designed for high load capacity. It can be easily switched to an omni-directional configuration using mecanum wheels. It features an IMU and can receive a camera and a laser scanner. It also features a default radio system for remote operation and is suitable for research and surveillance. It is controlled by a PC and it is programmed with open ROS architecture. Robotnik produces a line of industrial-grade robots, of which the Summit-XL is a midrange example.

Beside the platforms of Table 1, a reference should be given to a slightly different flavor of educational solutions, namely the kits by Vex Robotics [53] and Pitsco/Tetrix Robotics [54]. These kits provide robust metal parts, sensors, motors, electronics and other hardware for the assembly of a range of robots for education, hobby and competitions. They represent an advanced constructivist approach, with an average cost of a medium range kit of the order of 900 €.

The platforms presented above give a review of current educational robotic technology and trace its future evolution. Table 1 reveals a gap in low-cost educational platforms based on ROS. However, a unifying middleware like ROS is imperative for flexibility, adaptability, ease of development and community support. In addition, the above analysis shows that connectivity within a local computer network and browser-based programming tools are definite trends. Finally, the success of educational platforms depends on their low-cost and on the versatility of programming tools, from block-based to text-based programming, covering different educational levels and needs. These virtues were exploited in the design and implementation of the EUROPA platform.

3. Materials and Methods: Presentation of the EUROPA Platform

3.1. Overview of EUROPA

EUROPA (EdUcational Ros rObot PlAtform) is a two-wheel, inexpensive differential drive robot with a manipulator. It is adequately scalable and flexible to fit into different educational levels and different curricula. It allows programming with introductory or more advanced tools, depending on educational level. Its main controller is the Raspberry Pi 3 B+. An introductory presentation of the initial version of EUROPA was given in [51]. Figure 2 shows the basic EUROPA components.

EUROPA follows the open hardware paradigm and uses open source software. The robot can be built by the students themselves, under the appropriate instructions from their teachers, providing an opportunity for hands-on experience with principles of electricity, electronics and engineering. Although the robot can be used for Science, Technology, Mechanics and Mathematics (STEM) [2,3], it can also be upgraded with sensors like a LIDAR, to allow for more advanced lessons and research on robotics. EUROPA is based on ROS, which provides interoperability and extensibility. Although ROS stands for Robot Operating System, it is really a framework that sits on top of an existing operating system such as GNU/Linux. EUROPA includes a camera that can be used for image processing and object recognition. In addition, it supports a plethora of sensors that can be added to the Raspberry Pi board in order to support user-defined tasks.

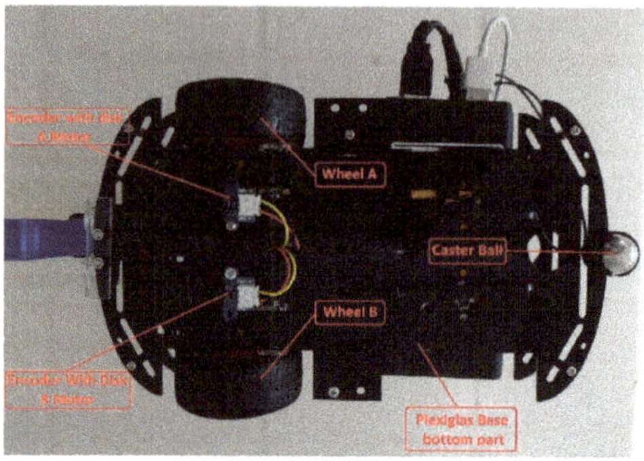

(a)

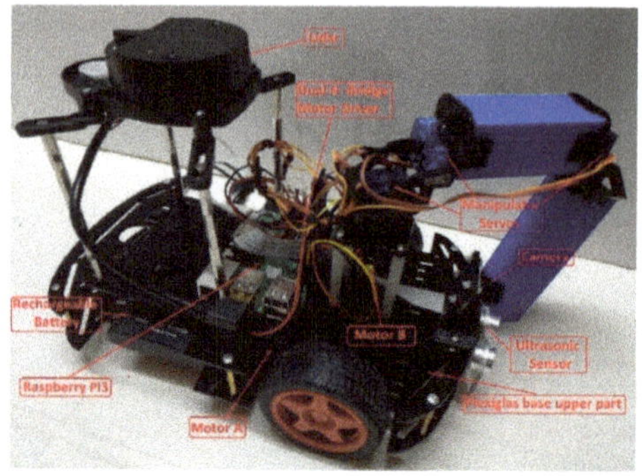

(b)

Figure 2. EUROPA and its components; (**a**) Bottom view; (**b**) Side view.

EUROPA includes a simulation environment. The robot was described in Unified Robot Description Format (URDF) and is simulated in the Gazebo environment [23]. Robot simulation allows children to easily and safely experiment with algorithms and develop skills related to computer programming. Following simulation, students will be able to choose the best performing algorithms, test them on the real robot in the physical world and understand the differences between robot behavior in a simulated as opposed to a real environment. In addition, using the interface of rviz [25], the popular visualization tool for ROS, they will be able to visualize depictions of the robot movement and easily control the robotic arm. Finally, they can learn concepts like odometry and sensor visualization.

3.2. EUROPA Hardware

EUROPA is built on a double plexiglas base, which supports all the robot's mechanical and electronic components. A rechargeable 10000 mAh battery is included, providing power to the Raspberry Pi and motors. Two differentially driven DC motors with wheels and encoder disks are responsible for EUROPA's locomotion, allowing a speed of up to 2 m/s with 8 N cm of maximum torque. This is enough for climbing on small ramps. In addition to the wheels, the robot rests on an omnidirectional caster ball, located on the back.

On the upper side, we find the Raspberry Pi 3 B+ board, a two-motor controller shield dual H-bridge motor driver DRV8833 [55], the Raspberry Pi Camera Module Night Vision-Adjustable Focus (5MP, 1080p) [56] and the robotic arm. The arm rests on a base made of 4 spacers 5 cm long screwed directly onto the robot chassis. The two axes of the arm are 3D printed and the joints are two Mini Pan-Tilt Kits powered by micro servo motors (Servo Micro plastic gears Feetech FS90, 1.5 kg·cm). The whole construction is characterized by simplicity and ease of assembly.

The University Edition of EUROPA features a laser scanner for 360 degrees distance measurement (LIDAR LDS1.5 [57]). It can measure a cloud of data around the robot up to a distance of 3.5 m and can support experiments on Simultaneous Localization and Mapping (SLAM).

At the bottom of the chassis there are two encoders with led-photodiode pairs (Waveshare, 12225) [58] used for odometry measurements. Finally, there is a distance meter on the front of the robot that can be used for obstacle avoidance (Ultrasonic Sensor 2–400 cm SR04) [59]. All accessories are connected directly to the Raspberry board without the need for extra electronic controllers. Table 2 presents the parts list and their approximate costs.

Table 2. List of EUROPA parts and approximate cost.

Part	Number	Cost in Euros
Raspberry 3 B+ board	1	41.90
Motor controller shield DRV8833	1	5.20
Raspberry Pi camera	1	21.20
Servo motors Feetech FS90	3	3 × 2.5 = 7.50
Led photodiode encoders	2	2 × 3.90 = 7.80
Ultrasonic sensor	1	2.50
DC motors with encoder disks	2	2 × 1.8 = 3.60
Plexiglas double base	1	1.80
Caster ball	1	1.60
Rechargeable battery	1	9.90
Wheels	2	2 × 1.5 = 3.00
Cables		4.00
Lidar	1	180.9
mini pan tilt kits	2	2 × 2.9
robotic arm axles (3d printed)	2	0.5
Spacers, bolts, nuts		2.0
Total Cost		299.2

3.3. EUROPA Software

EUROPA uses ROS infrastructure for communication and control, as shown in Figure 3. Raspberry controls the motors and is in charge of data collection from sensors and the camera. All the drivers responsible for the control of the two DC motors, the servo motors, the ultrasonic sensor and the LIDAR are installed on Raspberry Pi. The Raspberry Pi hosts several Python scripts that act as ROS nodes. For example, they collect video from the camera [60], receive input from the LIDAR [61], measure wheel movement via wheel encoders to calculate odometry and publish the data as ROS topics. A desktop computer which is running the ROS master is connected to the robot via Wi-Fi. Using the computer, the student or teacher can run either Python scripts or ROS user interfaces (UIs) [62] to control the movement of the robot and visualize data. The robotic arm can also be controlled via rviz or RQT [24,25] from the computer. Additionally, the robot can be controlled by a mobile phone, using ROS Control API without the need of a computer. Although the proposed way is to work with the robot from a remote computer, the student or instructor can also connect a screen and a keyboard directly to the Raspberry Pi and control the robot without the need of any additional device. The rviz-based user interface can also show live video from the camera and data from the ultrasonic sensor.

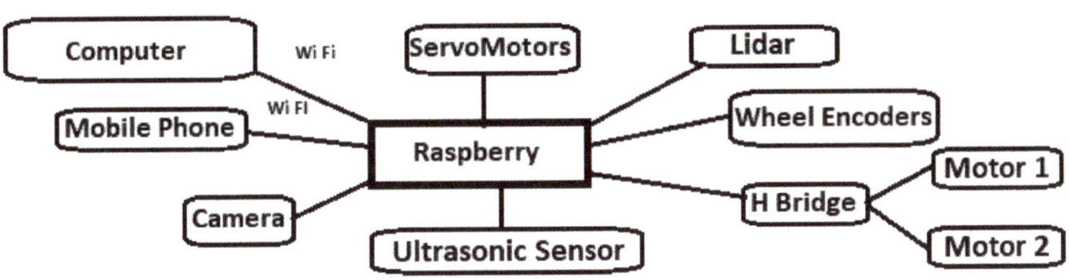

Figure 3. EUROPA control structure.

Additionally, the stream from the camera can be used to devise solutions to problems such as line following, while LIDAR and odometry can be used for Simultaneous Localization and Mapping (SLAM) [63] and navigation.

ROS and Gazebo provide communication, simulation and visualization tools. These modules are needed to perform tasks such as image processing and sensor calibration. Robot Operating System allows the use of modules and applications available in the ROS ecosystem. ROS provides modules for navigation, arm manipulation and SLAM. The ROS master is running on the PC, which is responsible for the communication between the various nodes running on the robot and computer. Any nodes that run in the ROS cluster can communicate with each other by exchanging information. Information circulates in the form of messages organized in topics, to which each one node either publishes or subscribes.

The different nodes that exist in EUROPA robot are described below. All nodes referred to as custom nodes have been created by the EUROPA team for use with the EUROPA robot.

3.3.1. Nodes that Run Exclusively on the Robot

- A custom Python node for DC motor control.
- A custom Python node for translating position information to appropriate command signals for the servo motors of the robotic arm.
- A node for streaming video from the camera.
- A node for the LIDAR operation which publishes data using ROS hls_lfcd_lds_driver
- driver [61].
- A custom Python node for publishing distance measurements collected from the distance sensor.
- A custom Python node for publishing odometry data from the wheel encoders.

3.3.2. Nodes that Run either on the Remote Computer or on the Robot itself (If It is Connected to a Screen and a Keyboard)

- A custom Python node for moving the robot using the keyboard.
- A custom Python node for the movement of the robotic arm.
- A custom node to watch the video captured by the robot camera
- A custom Python node to identify color lines and to send velocity messages that control the movement of the robot.
- A custom Python node for moving the robot to a specific position on the xy plane.

3.3.3. Nodes and Simulations that Run Exclusively on the Remote Computer

- Simulation of the robot in a virtual environment via the Gazebo application.
- Visualization of the robot movements and odometry via rviz.
- Control of the movements of the robotic arm through Moveit [64].
- A node responsible for SLAM using ROS's hector_slam [65] metapackage.

4. Results and Discussion: EUROPA in the Real World

4.1. Europa in Secondary Education

Most approaches to school robotics are currently focused on writing a script of code for robot control, along with a Lego-type construction. Usually, students do not go deeper into hardware and seldom do they go properly into software design concepts. The complexity of issues like motor control, wheel encoders and other sensors is usually hidden even from the interested student. One goal of the EUROPA project is to provide the students with an open platform for mechatronics concepts, ranging from introductory to advanced. The teacher can choose to present a high-level overview of the system or to teach in depth concepts. The students can acquire hands on experience with experiments in physics, electricity and robotics. EUROPA was tested in two Greek schools, during the first semester of the school year 2019-20. A STEM curriculum with applications in sciences, engineering and programming was designed and implemented. The target group was second-grade high school students, in the Greek system, which is equivalent to tenth or eleventh grade in the K12 system (ages 16–17). The curriculum that was used is briefly described below.

4.1.1. Robot Construction

The robot was constructed by the students with instructions from the teacher, and at the same time, an introductory lesson on sensors and motors was given. Initially, there was a reference to voltage, current and operation of DC motors. Past school lessons on these topics were revisited.

4.1.2. Motors and Sensors

The next step was to provide students with a basic understanding of the role of sensors and actuators. A presentation was given on servo motors and Pulse Width Modulation (PWM) was explained. A lesson on sensors was given, and different sensors were presented. The principle of data acquisition in a digital system was introduced and the use of a library for transferring data from a sensor to a Python program was explained. Then, the principle of the distance sensor was illustrated using a simple setup with a speaker and a microphone. Students were asked to calculate distance using time of flight, revisiting first grade physics. The photo-interrupter included in EUROPA provided the opportunity to introduce aspects of the interaction of light with matter. Finally, the camera was introduced and a reference to image processing was made. The role of the camera in the recognition of the environment was discussed.

4.1.3. Robot Simulation

In the next lesson, the robot's simulation was presented to students using rviz [37] and Gazebo environments. The students were also given the Unified Robot Description Format (URDF) file describing the robot. The XML file was analyzed focusing on specific physical properties of the robot. The students understood how a robot can be described using geometric figures and physical properties. Then, the students experimented by changing specific parameters to the existing robot description and saw how the robot was affected in the virtual environment.

4.1.4. Writing Python Scripts for EUROPA (Part 1)

The next lesson presented a Python script that receives input from the computer keyboard and translates it into robot motion commands. The students applied knowledge from lessons on circular motion and revisited notions on angular and linear velocity, applying them in real-world conditions.

4.1.5. Writing Python Scripts for EUROPA (Part 2)

The next lesson was to direct the robot to a specific position by applying the Pythagorean Theorem and other basic trigonometric equations. A Python script was created and explained before execution. At this point, it is important to note that students were watching the robot movements both in the simulation environment and in real life.

4.1.6. Data Acquisition from Wheel Encoders and Odometry Computation

During this lesson, the students first learned to use interrupts in order to get the encoder data and thus calculate angular velocity of each wheel. Additionally, students calculated odometry by applying high school grade physics kinematics and published odometry information to ROS.

4.1.7. Controlling the Robot Arm of EUROPA (Part 1)

The next lesson focused on the robotic arm of the robot. Geometry and algebra were linked to the movement of the arm. In addition, students were introduced to the concept of torque and they were familiarized with it by using different gears in Lego constructions. Continuing with this lesson, a simple movement of the arm was performed. A Python script for arm control was provided, and the students were asked to parameterize it. In addition, they used rviz with RQT to control the robotic arm. Different angles for the servomotors were given, and the students tried to determine theoretically the position of the tip of the robotic arm.

4.1.8. Controlling the Robot Arm of EUROPA (Part 2)

Next, the students were asked to calculate the angles of the servo motors of the robotic arm in order to place the tip at a specific position in 3D space. In this way, they were introduced to the importance and difficulty of the inverse kinematic problem. When they understood the difficulty of the problem, the Moveit! package [64] was presented, which provides the arm with the capability to perform complex movements using ready-made libraries and kinematic model solutions. Again, the students had the opportunity to see the simulated and real robot repeating the same movements.

4.2. *Advanced Robotics Course with EUROPA*

With the addition of the camera and the LIDAR, EUROPA becomes an efficient platform for teaching more advanced robotics courses. Such courses are often part of the curriculum in college or university; however, interested high school students can be benefited as well. After a series of introductory notions, students can continue the learning process, focusing on concepts related to computer vision, machine learning and robot autonomy. The following experiments were demonstrated in the same class of high school students who attended the set of lessons outlined in paragraph 4.1.

4.2.1. Tele-Operation of EUROPA Using the Camera

This project includes tele-operation of the robot using the camera and the distance meter. Students were viewing live video from the robot's camera displayed on their computer, and through this image they tele-operated the robot from their computer keyboard. To improve the movements of the robot in the room, they also used distance measurement and a simple obstacle avoidance Python script.

4.2.2. Line Following Using the Camera

The goal of this project was to use the camera as a color sensor in order to direct EUROPA to follow a yellow line painted on the floor. At the beginning of the lesson, the principles of digital vision sensors were explained to students and a reference was made to RGB color space. A simple experiment with the camera and a simple user interface based on OpenCV demonstrates how object colors are transformed in RGB values, using the camera. Then, HSV color space was introduced and an explanation was given as to why it is best to use HSV in conditions of unstable luminosity. A simple line-following algorithm based on color detection was presented and was applied in a Python script based on OpenCV. The program measures deviations from the yellow line measured in pixels and transforms them into appropriate wheel speeds for the differential drive. The algorithm is robust and results in smooth line following, better than using the infrared sensor commonly applied to this kind of experiment.

4.2.3. Simultaneous Localization and Mapping

The last experiment introduces the advanced topic of Simultaneous Localization and Mapping (SLAM). At the beginning of the lesson, a reference was made to how the LIDAR works, and the students saw a point cloud in rviz, representing the distances from the obstacles in the room. Subsequently, reference was made to mapping and its importance in robotics. Finally, the concept of Bayesian update using sensor measurements was introduced in general terms and a connection was made to similar concepts taught in mathematics lessons on probabilities. The Hector SLAM function [65] was introduced and students saw the mapping of their classroom in rviz.

4.3. *Performance Evaluation Experiments*

4.3.1. Odometry Evaluation

During this test, we commanded the robot to traverse a predefined orthogonal path with dimensions 0.8 m by 2 m and return to its original position. In Figure 4, the performed path is shown in red. The robot follows the commands quite accurately, with a most notable deviation from the commanded path observed during the final stages of the route. Although the robot was commanded to end up exactly at its starting point, the difference between the end point and the starting position is 3 cm in the horizontal axis and 9 cm in the vertical one. The blue line corresponds to the odometry as perceived by the robot. Odometry was measured using optical encoders. The axes units in the figure correspond to cm.

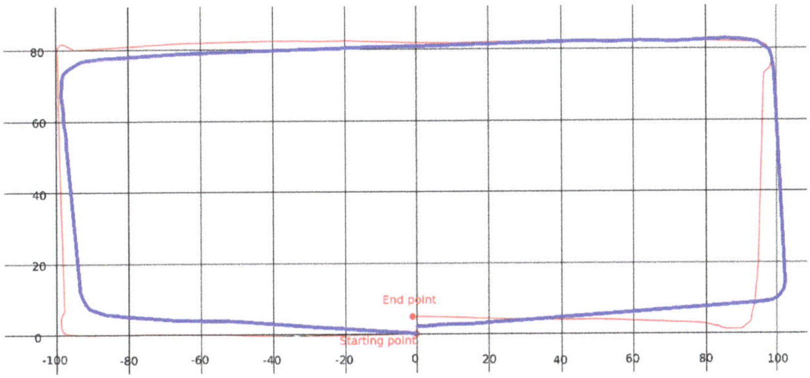

Figure 4. Evaluation of odometry measurements using the optical encoders.

4.3.2. Mapping

Hector-SLAM is a ROS package that is able to solve the robot localization and mapping problem for a 6DOF vehicle equipped with a laser scanning system (LIDAR) and inertial sensors [66]. The package fuses 3D robot attitude and position information obtained from an Inertial Measurement Unit with a 2D SLAM process. The SLAM process in Hector-SLAM is based on occupancy grid mapping combined with 2D pose estimation. At each step, the system aligns the new laser scan endpoints with the map learned so far. The optimization of the alignment process results in an estimate of the new position of the robot in the 2D map. In this way, the environmental map and the robot pose are produced incrementally, starting from a known pose.

In our EUROPA robot, the movement is on a plane and only the position (x, y) and orientation ψ on the plane is relevant. Therefore, only the 2D SLAM process is active and IMU information is not required. We have built a model environment, with approximate dimensions 2 m × 2 m, which can be traversed by the robot, starting from a known initial position and completing full circles around a corridor. At each step, the laser scanner acquires a cloud of points from the surrounding walls and computes the new change in translation and orientation, based on a transformation that gives the best alignment with the previous map. Knowing the new pose, the occupancy grid is updated.

In Figure 5 the environmental map created by the Hector-SLAM process is presented. Occupied cells are shown in black, while lighter color represents empty space. The red line illustrates the ground truth information of the model environment. The green line is the pose as it is computed during the SLAM process. The mapping was created after two loops around the corridor and the vehicle ended at its starting position. The Hector-SLAM package does not require odometry from optical encoders as input in the process.

The origin of the inertial frame is considered to coincide with the starting point of the robot track. The map was generated by teleoperating the robot with linear velocities less than 0.5 m/sec and angular velocities less than 0.314 rad/sec. Both the map and the pose estimation are considered to be satisfactory for the educational purpose served by our experiment.

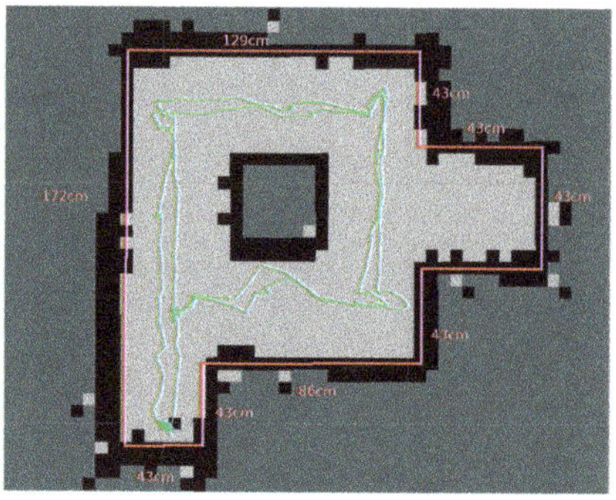

Figure 5. Evaluation of the robot's ability to create maps.

4.3.3. Line Following Performance

In order to assess the performance of the line following problem, we have created a yellow curved path of a total of 4.2 m, as in Figure 6.

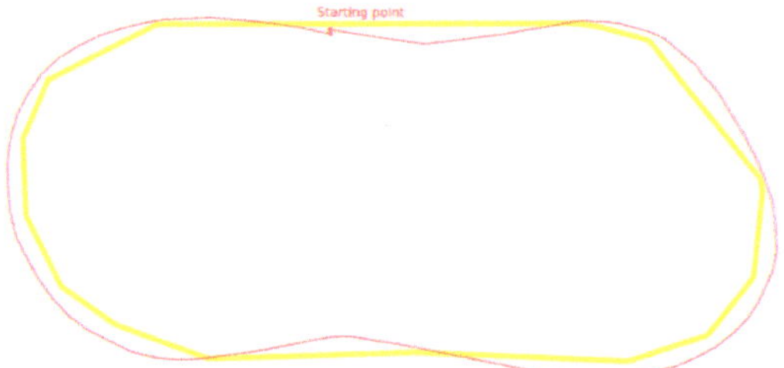

Figure 6. Line-following setup—The yellow line is the target path, the red line is the true path.

The actual robot path is shown by the red line. In Figure 7, the displacement of the robot from the center of the yellow line is shown, measured in mm. In this figure, the horizontal axis is the distance covered by the robot. At the starting point the robot was not in the center of the yellow line and had a 10-degree clockwise rotation. The line-following algorithm extracted color features from the image frame captured by the camera, using the OpenCV library. A simple P-controller was selected in order to direct the robot across its path. The P-controller was selected for educational reasons. The oscillations observed in Figure 7 are mainly due to the simplicity of the controller.

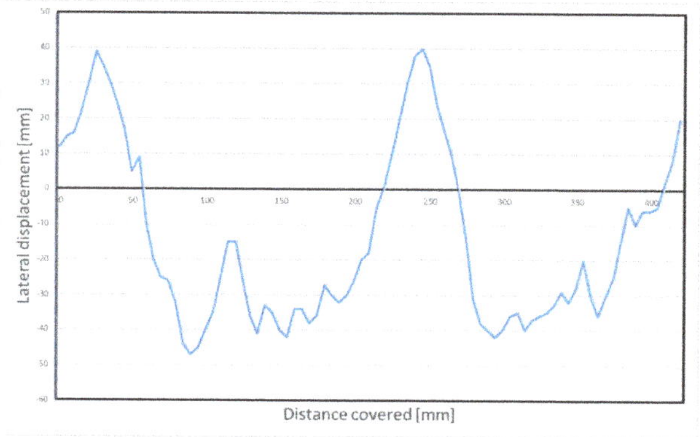

Figure 7. Line-following—displacement of the robot related to the line.

4.4. Assessment of EUROPA in the Classroom

The target group for the assessment was two second-grade high school classes or 22 and 23 students in Greece of ages 16 to 17. The lessons were performed as a part of a series of technology projects that have been added into the Greek curriculum during the past few years and correspond to a weekly workload of 2 h. The main goal of EUROPA in these technology projects was to provide students with real world science examples and a better understanding of notions that they have already been taught in lessons such as physics, mathematics and computer science. The students were acquainted with more advanced technological subjects and were motivated for independent learning and discovery. The acceptance of the platform was enthusiastic. All students were able to follow, understand and work on the EUROPA robots without any serious problems and some of them were even willing to drill down to the robot's architecture.

Regarding physics and mathematics, EUROPA's impact was evident since among others, students had a chance to relate theoretical kinematics and dynamics with practical robot movement. They saw that the distribution of mass in an object can affect its movement. They applied theoretical knowledge on rotational movement to wheel rotation and connected it to odometry calculations. They also had a chance to apply trigonometry and vector analysis to real world problems. EUROPA also proved to be a great medium for the introduction of students to new concepts such as sensors, actuators, control, and physical computing. In programming lessons, the students applied programming skills in solving real problems, which gave them a totally new incentive for writing code and understanding programming structures. They faced the notion that hardware abstraction and standards are particularly important in order to make sensory information usable and that working in simulation is quite different than working in the real world.

At the end of the semester, the students had clearly a better understanding of real-world problems solved by science, and their interest in technology was higher than with a similar course designed with a LEGO-like platform. The openness of the platform and the advanced scenarios that were shown to the children proved to be quite important to motivate them.

5. Conclusions

This paper reviews existing educational robot platforms for various levels of education, from kindergarten to university. Our research reveals that there is a gap in the low-cost range of ROS-based educational robotics. However, ROS-based robotics is versatile and has great potential for integration with free simulation and visualization software, as well as with advanced sensors. A ROS-based, low-cost platform can support advanced projects, like machine vision, machine intelligence, localization and mapping. This gave us the incentive to build EUROPA which is a cheap and versatile open platform based on ROS. It can cover a range of applications, from basic educational robotics to advanced applications, such as vision and mapping. Its main controller is the Raspberry Pi, which is supported by a great community and can readily use a plethora of applications. The platform is currently being assessed in two secondary schools in Central Macedonia, Greece, under a pilot robotics curriculum. Future work includes redesigning both the platform and the curriculum, after receiving feedback from pilot schools. We also aim to build an online community, supporting students and teachers with educational material and extensive documentation.

Author Contributions: Conceptualization, S.V., G.K. and J.K.; methodology, S.V. and G.K.; software, G.K.; validation, G.K., S.V. and J.K.; investigation, G.K. and J.K.; resources, J.K.; writing—original draft preparation, G.K.; writing—review and editing, S.V. and J.K.; supervision, S.V.; project administration, J.K.; All authors have read and agreed to the published version of the manuscript.

Funding: This research received no external funding.

Conflicts of Interest: The authors declare no conflict of interest.

References

1. Papert, S. *Mindstorms: Children, Computers, and Powerful Ideas*; Basic Books: New York, NY, USA, 1980.
2. Benitti, F.B.V. Exploring the educational potential of robotics in schools: A systematic review. *Comput. Educ.* **2012**, *58*, 978–988. [CrossRef]
3. Benitti, F.B.V.; Spolaôr, N. How have robots supported STEM teaching? In *Robotics in STEM Education*; Khine, M.S., Ed.; Springer: Cham, Switzerland, 2017; pp. 103–129.
4. Jeon, M.; Fakhrhosseini, M.; Barnes, J.; Duford, Z.; Zhang, R.; Ryan, J.; Vasey, E. Making live theatre with multiple robots as actors. In Proceedings of the 11th ACM/IEEE International Conference on Human-Robot Interaction (HRI 2016), Christchurch, New Zealand, 7–10 March 2016; pp. 445–446.
5. Jeon, M.; Barnes, J.; Fakhrhosseini, M.; Vasey, E.; Duford, Z.; Zheng, Z.; Dare, E. Robot opera: A modularized afterschool program for STEAM education at local elementary school. In Proceedings of the 14th International Conference on Ubiquitous Robots and Ambient Intelligence (URAI 2017), Jehu, Korea, 28 June–1 July 2017; pp. 935–936.
6. Zhang, Z.; Zhang, A.; Zhang, M.; Esche, S. Project-based Courses for B.Tech. Program of Robotics in Mechanical Engineering Technology. *ASEE Comput. Educ. (CoED) J.* **2020**, *11*, 1–10.
7. Sullivan, A.; Bers, M.U. Robotics in the early childhood classroom: Learning outcomes from an 8-week robotics curriculum in pre-kindergarten through second grade. *Int. J. Technol. Des. Educ.* **2015**, *26*, 3–20. [CrossRef]
8. Elkin, M.; Sullivan, A.; Bers, M.U. Programming with the KIBO robotics kit in preschool classrooms. *Comput. Sch.* **2016**, *33*, 169–186. [CrossRef]
9. Huskens, B.; Palmen, A.; Van der Werff, M.; Lourens, T.; Barakova, E.I. Improving collaborative play between children with autism spectrum disorders and their siblings: The effectiveness of a robot-mediated intervention based on Lego® therapy. *J. Autism Dev. Disord.* **2015**, *45*, 3746–3755. [CrossRef] [PubMed]
10. Lytridis, C.; Bazinas, C.; Papakostas, G.A.; Kaburlasos, V. On Measuring Engagement Level During Child-Robot Interaction in Education. In *Advances in Intelligent Systems and Computing, Proceedings of the Robotics in Education (RiE 2019), Vienna, Austria, 10–12 April 2019*; Springer: Cham, Switzerland, 2019; Volume 1023.
11. Merdan, M.; Lepuschitz, W.; Koppensteiner, G.; Balogh, R.; Obdržálek, D. (Eds.) *Advances in Intelligent Systems and Computing, Proceedings of Robotics in Education (RiE 2019), Vienna, Austria, 10–12 April 2019*; Springer: Cham, Switzerland, 2020; Volume 1023.
12. Alimisis, D.; Moro, M. (Eds.) *Special Issue on Educational Robotics, Robotics and Autonomous Systems*; Elsevier: New York, NY, USA, 2016; Volume 77, pp. 1–76.
13. Paull, L.; Tani, J.; Ahn, H.; Alonso-Mora, J.; Carlone, L.; Cap, M.; Chen, Y.F.; Choi, C.; Dusek, J.; Fang, Y.; et al. Duckietown: An Open, Inexpensive and Flexible Platform for Autonomy Education and Research. In Proceedings of the IEEE International Conference on Robotics and Automation (ICRA 2017), Singapore, 29 May–3 June 2017; pp. 1497–1504.
14. ROS Wiki, Turtlebot 3. Available online: https://wiki.ros.org/Robots/TurtleBot (accessed on 6 December 2019).
15. Pedersen, B.K.M.K.; Larsen, J.C.; Nielsen, J. The Effect of Commercially Available Educational Robotics: A Systematic Review. In *Advances in Intelligent Systems and Computing, Proceedings of Robotics in Education (RiE 2019), Vienna, Austria, 10–12 April 2019*; Merdan, M., Lepuschitz, W., Koppensteiner, G., Balogh, R., Obdržálek, D., Eds.; Springer: Cham, Switzerland, 2020; Volume 1023, pp. 14–27.
16. Piaget, J.; Inhelder, B. *The Child's Construction of Quantities*; Routledge and Kegan Paul: London, UK, 1974.
17. Papert, S. *The Children's Machine—Rethinking School in the Age of the Computer*; Basic Books: New York, NY, USA, 1993; ISBN 13: 978-0465010639.
18. Karim, M.E.; Lemaignan, S.; Mondada, F. A review: Can robots reshape K-12 STEM education? In Proceedings of the IEEE International Workshop on Advanced Robotics and its Social Impacts, Lyon, France, 1–3 July 2015; pp. 1–8.
19. Kushner, D. The Making of Arduino. Available online: http://spectrum.ieee.org/geek-life/hands-on/the-making-of-arduino (accessed on 6 December 2019).
20. Arduino Web Site. Available online: https://www.arduino.cc/ (accessed on 6 December 2019).
21. Raspberry Foundation Web Page. Available online: https://www.raspberrypi.org/ (accessed on 6 December 2019).
22. Quigley, M.; Conley, K.; Gerkey, B.; Faust, J.; Foote, T.B.; Leibs, J.; Wheeler, R.; Ng, A.Y. ROS: An open-source robot operating system. In *ICRA Open-Source Software Workshop*; ICRA: Kobe, Japan, 2009.
23. Gazebo Simulator Web Page. Available online: http://gazebosim.org/ (accessed on 6 December 2019).

24. Rqt rviz, Package Summary. Available online: http://wiki.ros.org/rqt_rviz (accessed on 6 December 2019).
25. Rviz—3D Visualization Tool for ROS. Available online: http://wiki.ros.org/rviz (accessed on 6 December 2019).
26. Roussou, E.; Rangoussi, M. On the Use of Robotics for the Development of Computational Thinking in Kindergarten: Educational Intervention and Evaluation. In *Robotics in Education, RiE 2019. Advances in Intelligent Systems and Computing*; Merdan, M., Lepuschitz, W., Koppensteiner, G., Balogh, R., Obdržálek, D., Eds.; Springer: Cham, Switzerland, 2020; Volume 1023, pp. 31–44.
27. Highfield, K. Robotic toys as a catalyst for mathematical problem solving. *Aust. Primary Math. Classroom* **2010**, *15*, 22–27.
28. Thymio II Web. Available online: https://www.thymio.org/home-en:home (accessed on 6 December 2019).
29. Mondada, F.; Bonani, M.; Riedo, F.; Briod, M.; Pereyre, L.; Retornaz, P.; Magnenat, S. The Thymio open-source hardware robot. *IEEE Robot. Autom. Mag.* **2017**, *1070*, 77–85. [CrossRef]
30. Parallax Web Page. Available online: https://www.parallax.com (accessed on 6 December 2019).
31. Souza, I.; Andrade, W.; Sampaio, L.; Araújo, A.L.; Araujo, S. A Systematic Review on the use of LEGO® Robotics in Education. 2018. Available online: https://www.researchgate.net/publication/328410916 (accessed on 6 December 2019).
32. Lego Web Page. Available online: https://www.lego.com (accessed on 6 December 2019).
33. Casini, M.; Garulli, A.; Giannitrapani, A.; Vicino, A. A Remote Lab for Experiments with a Team of Mobile Robots. *Sensors* **2014**, *14*, 16486–16507. [CrossRef] [PubMed]
34. Mondada, F.; Bonani, M.; Raemy, X.; Pugh, J.; Cianci, C.; Klaptocz, A.; Magnenat, S.; Zufferey, J.-C.; Floreano, D.; Martinoli, A. The e-puck, a Robot Designed for Education in Engineering. In Proceedings of the 9th Conference on Autonomous Robot Systems and Competitions, Castelo Branco, Portugal, 7 May 2009; Volume 1, pp. 59–65.
35. Turtle-bot Web Page. Available online: http://www.robotis.us/turtlebot/ (accessed on 6 December 2019).
36. Duckietown Web Page. Available online: https://www.duckietown.org/ (accessed on 6 December 2019).
37. Navaro, P.J.; Fernandez, C.; Sanchez, P. Industrial-like Vehicle Platforms for Postgraduate Laboratory Courses on Robotics. *IEEE Trans. Educ.* **2013**, *56*, 34–41. [CrossRef]
38. Leo Web Page. Available online: https://www.leorover.tech/ (accessed on 6 December 2019).
39. Robotnik Web Page. Available online: https://www.robotnik.eu/mobile-robots/summit-xl-hl/ (accessed on 6 December 2019).
40. Clearpath Web Page. Available online: https://clearpathrobotics.com/robots/ (accessed on 6 December 2019).
41. SoftBank Web Page. Available online: https://www.softbankrobotics.com/emea/index.php/en/nao (accessed on 6 December 2019).
42. Humanizing Technologies Web Page. Available online: https://www.humanizing.com/our-robots/ (accessed on 6 December 2019).
43. Robotis OP3 Web Page. Available online: http://www.robotis.us/robotis-OP2-OP3/ (accessed on 6 December 2019).
44. Makeblock Mbot Web Page. Available online: https://www.makeblock.com/steam-kits/mbot (accessed on 6 December 2019).
45. Magnenat, S.; Mondada, F. ASEBA Meets D-Bus: From the Depths of a Low-Level Event-Based Architecture into the Middleware Realm. In Proceedings of the IEEE TC-Soft Workshop on Event-based Systems in Robotics (EBS-RO), St. Louis, MO, USA, 15 October 2009.
46. Edison Web. Available online: https://meetedison.com/ (accessed on 6 December 2019).
47. Waveshare Web. Available online: https://www.waveshare.com/wiki/AlphaBot2 (accessed on 6 December 2019).
48. Bellas, F.; Mallo, A.; Naya, M.; Souto, D.; Deibe, A.; Prieto, A.; Duro, R.J. STEAM Approach to Autonomous Robotics Curriculum for High School Using the Robobo Robot. In *Robotics in Education, RiE 2019. Advances in Intelligent Systems and Computing*; Merdan, M., Lepuschitz, W., Koppensteiner, G., et al., Eds.; Springer: Cham, Switzerland, 2020; Volume 1023, pp. 77–89.
49. Naya, M.; Varela, G.; Llamas, L.; Bautista, M.; Becerra, J.C.; Bellas, F.; Prieto, A.; Deibe, A.; Duro, R. A Versatile Robotic Platform for Educational Interaction. In Proceedings of the 9th International Conference on Intelligent Data Acquisition and Advanced Computing Systems: Technology and Applications, Bucharest, Romania, 21–13 September 2017.
50. Robobo Web Page. Available online: https://theroboboproject.com/en/ (accessed on 6 December 2019).

51. Karalekas, G.; Vologiannidis, S.; Kalomiros, J. EUROPA—A ROS-based Open Platform for Educational Robotics. In Proceedings of the 10th IEEE International Conference on Intelligent Data Acquisition and Advanced Computing Systems: Technology and Applications, Metz, France, 18–21 September 2019.
52. Amsters, R.; Slaets, P. Turtlebot 3 as a Robotics Education Platform. In *Robotics in Education, RiE 2019. Advances in Intelligent Systems and Computing*; Merdan, M., Lepuschitz, W., Koppensteiner, G., Balogh, R., Obdržálek, D., Eds.; Springer: Cham, Switzerland, 2020; Volume 1023, pp. 170–181.
53. Vex Robotics Web Page. Available online: https://www.vexrobotics.com/ (accessed on 6 December 2019).
54. Pitsco/Tetrix Robotics Web Page. Available online: https://www.pitsco.com/Shop/TETRIX-Robotics (accessed on 6 December 2019).
55. Motor Driver Web Page and Datasheet. Available online: http://www.ti.com/lit/ds/symlink/drv8833.pdf (accessed on 6 December 2019).
56. Raspberry Pi Camera User Manual. Available online: https://www.waveshare.com/w/upload/6/61/RPi-Camera-User-Manual.pdf (accessed on 6 December 2019).
57. LDS-01 Web Page. Available online: http://www.robotis.us/360-laser-distance-sensor-lds-01-lidar/ (accessed on 6 December 2019).
58. Photo-Interrupter Sensor Wiki. Available online: https://www.waveshare.com/wiki/Photo_Interrupter_Sensor (accessed on 6 December 2019).
59. Ultrasonic Ranging Module HC-SR04. Available online: https://cdn.sparkfun.com/datasheets/Sensors/Proximity/HCSR04.pdf (accessed on 6 December 2019).
60. ROS Node for Camera Module of Raspberry Pi. Available online: https://github.com/UbiquityRobotics/raspicam_node (accessed on 6 December 2019).
61. ROS Wiki Web Page, LDS Driver. Available online: http://wiki.ros.org/hls_lfcd_lds_driver (accessed on 6 December 2019).
62. ROS Wiki Web Page, Keyboard Teleoperation Package Summary. Available online: http://wiki.ros.org/teleop_twist_keyboard (accessed on 6 December 2019).
63. Fernandez-Madrigal, J.-A.; Claraco, J.L. *Simultaneous Localization and Mapping for Mobile Robots: Introduction and Methods*; IGI Global: Hershey, PA, USA, 2013; ISBN 13: 978-1466621046.
64. Sucan, I.A.; Chitta, S.M. Available online: http://moveit.ros.org (accessed on 6 December 2019).
65. ROS Wiki, Hector Slam Documentation. Available online: http://wiki.ros.org/hector_slam (accessed on 6 December 2019).
66. Kohlbrecher, S.; Meyer, J.; von Stryk, O.; Klingauf, U. A Flexible and Scalable SLAM System with Full 3D Motion Estimation. In Proceedings of the IEEE International Symposium on Safety, Security and Rescue Robotics (SSRR), Kyoto, Japan, 1–5 November 2011.

sensors

MDPI

Article

Data-Driven Living Spaces' Heating Dynamics Modeling in Smart Buildings using Machine Learning-Based Identification

Roozbeh Sadeghian Broujeny *, Kurosh Madani, Abdennasser Chebira, Veronique Amarger and Laurent Hurtard

Université Paris-Est, LISSI Laboratory EA 3956, Senart-FB Institute of Technology, Campus de Senart, 36-37 Rue Charpak–F-77567 Lieusaint, France; madani@u-pec.fr (K.M.); chebira@u-pec.fr (A.C.); amarger@u-pec.fr (V.A.); hurtard@u-pec.fr (L.H.)

* Correspondence: r.sadeghian88@gmail.com

Received: 28 December 2019; Accepted: 13 February 2020; Published: 16 February 2020

check for updates

Abstract: Modeling and control of the heating feature of living spaces remain challenging tasks because of the intrinsic nonlinear nature of the involved processes as well as the strong nonlinearity of the entailed dynamic parameters in those processes. Although nowadays, adaptive heating controllers represent a crucial need for smart building energy management systems (SBEMS) as well as an appealing perspective for their effectiveness in optimizing energy efficiency, unfortunately, the leakage of models competent in handling the complexity of real living spaces' heating processes means the control strategies implemented in most SBEMSs are still conventional. Within this context and by considering that the living space's occupation rate (i.e., by users or residents) may affect the model and the issued heating control strategy of the concerned living space, we have investigated the design and implementation of a data-driven machine learning-based identification of the building's living space dynamic heating conduct, taking into account the occupancy (by the residents) of the heated space. In fact, the proposed modeling strategy takes advantage, on the one hand, of the forecasting capacity of the time-series of the nonlinear autoregressive exogenous (NARX) model, and on the other hand, from the multi-layer perceptron's (MLP) learning and generalization skills. The proposed approach has been implemented and applied for modeling the dynamic heating conduct of a real five-floor building's living spaces located at Senart Campus of University Paris-Est Créteil (UPEC), taking into account their occupancy (by users of this public building). The obtained results assessing the accuracy and addictiveness of the investigated hybrid machine learning-based approach are reported and discussed.

Keywords: system identification; smart building; artificial neural network; energy efficiency; black box modeling

1. Introduction and Related Works

In the context of the perspicacious decrease of fossil fuel resources and ongoing increase of energy consumption innate to the intensification of human urban activities during the last decades, the management of energy consumption in commercial and residential buildings has become a vital question. Regarding the works of [1] and [2], in the USA, the contribution of energy consumption in space heating was responsible for 43 percent in residential buildings in 2015, and in commercial buildings, this contribution was about 25 percent in 2012. This shows the huge slice of energy consumption related to space heating in the above-mentioned two sections. The recent enhancement in smart building energy management systems (SBEMSs) or smart building management systems (SBMS) by controlling and reducing the above-mentioned share of energy consumption is becoming the most efficient trend for facing energy consumption growth in residential and commercial buildings.

Smart building energy management systems (SBEMSs) in smart dwellings provide the inhabitants with advanced monitoring and control of the building's functions and a clever way to manage heavy power-consuming appliances (as heating devices) in order to achieve energy efficiency while optimizing and preserving the inhabitants' (or users') comfortable environment [3].

Although the sensors' quality and the technological features of the remote devices forming the physical part of the automated or smart buildings play an undeniable role in the performance of SBEMSs in optimizing the building's energy consumption, the primary inefficiency of such systems in declining energy consumption is related to the quality of the models that bear either the identification of the relationship between the building's behavior and the controller that hatches up the actions of implemented sensors and remote devices or to the excellence of the control strategy in charge of the building's behavior control. Thus, the identification and modeling of the building's operational dynamics remain key points in BMSs and especially in SBEMSs. On the other hand, the diversity of the involved factors (parameters) as well as their highly nonlinear variation make the identification and modeling of the dynamic behavior of a building a challenging task. Within this context and by considering that besides the living space's intrinsic structural features, the occupation of the living space (by users or residents) may affect the model of heating dynamics of the concerned living space, we have investigated the design, implementation, and validation of a data-driven machine learning-based identifier supplied by the time-series prediction paradigm's formalism. In fact, the human body continuously produces thermal energy, mostly in the form of heat radiation emission. Regarding black body law, a human in a sitting position and at about 1.80 m in height can emit 100 watts [4–6].

A number of works address model-free approaches coping with buildings' heating. Related to conventional controllers, the authors of [7] introduced a control heating system for supporting the heating comfort of the user based on a very simple thermostatic controller (operating on an "on/off" strategy) with the help of a microcontroller. When the temperature is higher than the desired temperature, the fan will turn on, and when the temperature is lower than the desired temperature, the heater will turn on. The proposed simplistic control of the space heating operates on the difference between the desired temperature and actual temperature, and could be seen as a model-free heating approach. While taking advantage of its independency from the effective complexity of the concerned edifice's hitting-dynamics, the proposed strategy is applicable to very specific homogenious living spaces, and cannot be generalized to more sophisticated buildings including heterogeneous living-spaces. In the work of [8], the investigator presents a gray-box methodology for thermal modelling of buildings. Gray-box modelling is a hybrid of data-driven and physics-based models, where coefficients of the equations from physics-based models are ajusted using data. The authors claim that the proposed methodology allows to capture the dynamics of the buildings while avoiding the effective complexity of the physics-based modelling, and results in simpler models. In fact, after first developing the individual components of the building such as temperature evolution, flow controller, and so on, the authors integrate these individual models into what they call the "complete gray-box model" of the building. The model has been validated using data collected from one of the buildings at Luleå, a city on the coast of northern Sweden. While using a simpler and generic model (compared with the physics-based complex heating models), the proposed approach remains far from convincing concerning its generalization to the other buildings.

The investigators of [9] propose a model-free and sensor-free *heating, ventilation and air-conditioning* (HVAC) control algorithm that uses simple user input (hot/cold) and adapts to changing office occupancy or ambient temperature in real time. As an alternative, the proposed strategy includes users in the HVAC control loop through distributed smart-phone based votes about their thermal comfort for aggregated control of HVAC. The developed iterative data fusion algorithm finds the optimal temperature in offices with multiple users and addresse techniques that can aggressively save energy by drifting indoor temperatures towards the outdoor temperature. The evaluation has been based on empirical data collected in 12 offices over a three-week period and showed that the proposed control may save up to 60% of energy at a relatively small increase in average occupant discomfort of 0.3 °C. While the idea is appealing, the concerned technique here also is very specific.

The control systems designed in [7–9] operate without any pre-knowledge of the living spaces that they are supposed to heat. In other words, the proposed solutions are based exclusively on data provided by temperature sensors within the frame of specific edifices for which the model of heating-dynamics is available. This makes the proposed models and issued controllers specific to the considered case studies, and thus not applicable to other structures (i.e., other buildings).

On the basis of the above-mentioned points, in the present article, we focus on the design and implementation of a data-driven machine learning-based identification of the building's living-space dynamic heating conduct, taking into account the occupancy (by the residents) of the heated space. This step is necessary for pulling off a comprehensive (i.e., interpretable) model handling the dynamic heating conduct of a living space with and without human presence. The proposed data-driven machine learning-based identifier will be applied for modeling the dynamic heating conduct of a real five-floor building's living spaces located at Senart Campus of University Paris-Est Créteil, taking into account their occupancy (by users of this public building).

From a general standpoint, identification approaches are divided into two main categories: white-box modeling (WBS) and black-box modeling (BBS) [10]. In WBS-based methods, the modeling of a system is performed on the basis of the formal relationship of the physical properties of the concerned system. If the main advantage of WBS-based methods remains their comprehensive and interpretable nature, however, often the effective complexity of real-world conditions causes WBS to lead to insolvable equations, and hence frequently to a strongly simplified issued model, making it quite far from the realistic behavior of the target system. In BBS-based methods, the modeling is done by mapping of an approximate behavior of the target system through the input–output relationship of that system. In contrast to WBS, if BBS-based methods achieve more accurate approximation of the effective complexity of the modeled system's behavior, often they lead to a shortfall of comprehensive and interpretable foundation related to the issued model.

Numerous research works have been accomplished in the past decades within the areas of identification and modeling of nonlinear systems related to our purpose. Wiener and Hammerstein-type models [11], Volterra series [12], and machine-learning based approaches such as fuzzy logic-based models [13] and artificial neural network-based approaches [14] have been presented. The authors of [15] identify a solar heating system utilizing BBS based on what they call the "recursive prediction error method" (RPEM). It is on the basis of a state-space model. The target system (namely a solar heater) includes two inputs (solar radiation energy and speed of the fan) and one output (air's temperature). They claim that the small amount of data necessary for the proposed approach is an advantage. However, the related simplicity of the target system and complicated expected behavior identification do not persuade the extendibility of the proposed approach to a realistic system including a large number of parameters (inputs and outputs).

In the work of [16], the identification of a heating system is done by investigation by means of an auto-regressive (ARX) model, auto-regressive and moving average (ARMAX) model, and Box–Jenkins (BJ) model. The target system includes a lamp and a metallic plate. It contains just one input (the lamp's voltage) and one output (the metallic plate's temperature). For the aforesaid case study, the authors used the system identification toolbox of MATLAB. However, the relative simplicity of the target system does not allow assessing the effectuality of the considered approach. It just presents that MATLAB's system identification toolbox is able to imitate this uncomplicated case study example. Similarly, the authors of [17] used MATLAB's identification toolbox for identification of the behavior of a boiler and heat exchanger transfer function. Nevertheless, the stated result does not end up with the accuracy of the target system identification. It results in a tough target device modeling. The authors of [18] provide the consequences of a dwelling's thermal model identification. It includes two bedrooms heated by electrical baseboard heaters. Owing to the modeling of the target system, the authors used EnergyPlus (software for simulating the building energy system providing functional modeling of energy consumption for heating, cooling, ventilation, and lighting in buildings). The control signal was simulated by MATLAB. The Building Controls Virtual Test Bed open-source software (of Berkley Lab. [19]) is a free, available co-simulation software linking different simulation programs as EnergyPlus, Modelica (an object-oriented language for complex systems' simulation [20]), and MATLAB/Simulink. In the account of the approximating dynamic of the system in Energy Plus, a low order state-space model is utilized. Concerning the identification of the system, they used N4SID subspace identification [21]. The authors in this investigation end up with a satisfactory average root-mean-square-error (RMSE) throughout ten reported simulated apartments. Nonetheless, they concluded that the time-consuming implementation makes it difficult to extend the proposed approach to more complicated systems.

The aforesaid investigations put emphasis on the pertinence of identification approaches for the modeling of buildings' heating dynamics. Indeed, all of the referenced investigations underline the tough limitations of the overviewed solutions in matching the complex behavior of space heating systems in buildings. The main shortages are either related to the eager simplification of the actual operative complexity of involved equations, in order to ease their computational solutions, or inherent to the nonlinearity and outsized number of the involved parameters. If the analysis of the aforementioned research works highlights the diversity of the covered fields and applications, they confirm what we mentioned before related to the advantages and shortages of each category (i.e., WBS-like and BBS-like) of identification-based nonlinear systems' modeling approaches. Meanwhile, the overviewed research works reveal the appealing capacity of the nonlinear autoregressive exogenous (NARX) model in modeling and forecasting complex systems' behaviors. In fact, the proposed modeling strategy takes advantage, on the one hand, from the forecasting capacity of the time-series of the NARX model, and on the other hand, from the multi-layer perceptron's (MLP) learning and generalization skills. If the NARX model has already been used for modeling in various paradigms, the originality of its application in the present article concerns its usage, and especially its closed-loop version, in the uninterrupted (i.e., continual) identification of the heating dynamics within a fully data-driven context. However, the additional novelties of the reported investigations, on the one hand, relate to the application of the aforementioned model for solving real-world problems addressing complex behaviors, and on the other hand, concern the effective implementation of the developed system by the use of standard technology (i.e., market available), overcoming complex technological obstacles.

Section 2 of this article presents the method and concepts of the proposed data-driven identification approach. Section 3 details the implementation of the issued method on SBEMS of the above-mentioned five-floor experimental building. The experimental setup, the experimental protocol, and the obtained results are presented and discussed. Finally, Section 4 concludes the article.

2. Machine Learning-Based Identification of the Heating Dynamics of the Living Space

Before bestowing the proposed living space heating dynamics identification approach, we consider the following work hypothesis relating to the identification strategy:

- The concerned living space is supposed to be part of a typical building including various quarters (such as flats and rooms for a residential building or working spaces, office rooms, classrooms, and practice rooms for a public building, and so on).
- The building is supposed to be heated by a central heater supplying radiators located in the aforementioned living spaces.
- The regulation is supposed to be done by a conventional controller adjusting the radiators' valves versus the magnitude of the outdoor temperature and the target (i.e., desired) indoor temperature.
- The target model considers the system to be identified as an overall system including the heat transmitters (radiators) and the heating space.

1. The concerned living space is supposed to contain an amount of N (with $0 \leq N \leq N_{Max}$) occupants (residents or users). $N = 0$ corresponds to an empty living space, while $N = N_{Max}$ characterizes a fully occupied living space. Thus, N_{Max} corresponds to maximum capacity of the living space and is determined according to the construction norms and occupation regulations.

Thus, within the aforementioned work hypothesis, the considered parameters are as follows: "Valve-position at time t" (denoted by $\vartheta_P(t)$), providing the heated water's flow (expressed as a normalized ratio of debit versus the maximum debit of the valve); "Outdoor-Temperature measured at time t" (denoted by $T_{Out}(t)$, expressed in °C); "Indoor-Temperature measured at time t" (denoted by $T_{In}(t)$, expressed in °C); and "Occupancy-Rate at time t" (denoted by $O_{CC}(t)$, expressed in %).

As mentioned in the introductory section, the identification method of the proposed system is accomplished by an MLP-based NARX ([22–24]) with a feed-forward back-propagation learning algorithm ([25,26]). Equation (1) and (2) respectively specify the overall open-loop NARX model, where $F(\cdot)$ is the activation function of the ANN, $\hat{y}(t+1)$ is the estimated (i.e., predicted) output, $y(t)$ is the actual output value of the model (i.e., at time t), $y\ (t-1), \cdots, y\ (t-n)$ are n-past values of the y(t), $x(t)$ is the present input value, and $y(t), y\ (t-1), \cdots, y\ (t-m)$ are the actual and tapped delayed exogenous inputs in m-past input values. Figure 1 illustrates the overall schema of the NARX model $\hat{T}_{In}(t)$.

$$\hat{y}(t+1) = F(y\ (t),\ y\ (t-1), \cdots, y\ (t-n),\ x\ (t),\ x\ (t-1),\ \cdots,\ x\ (t-m)) \quad (1)$$

$$\hat{y}(t+1) = F(\hat{y}(t),\ \hat{y}(t-1), \cdots, \hat{y}(t-n),\ x\ (t),\ x\ (t-1),\ \cdots,\ x\ (t-m)) \quad (2)$$

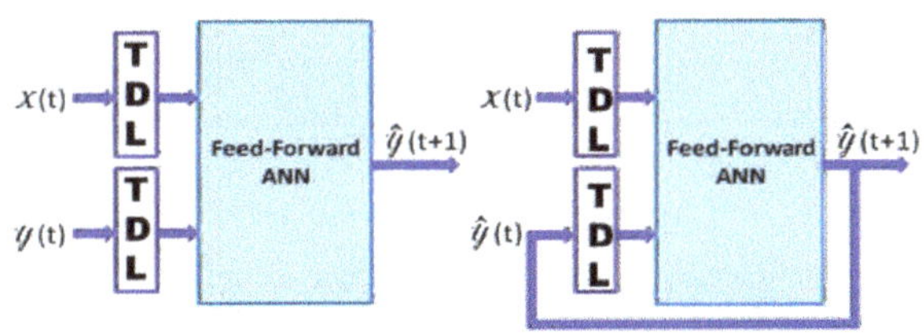

Figure 1. General artificial neural network (ANN)-based nonlinear autoregressive exogenous (NARX) model: open-loop (left) and closed-loop (right) architectures.

Figure 2 reveals the proposed identification structure of the target heating model, taking into account the above-stated work hypothesis including the influence of the occupancy. The learning process is performed by utilizing the open-loop NARX scheme. The learning dataset contains the operation of the real system's sequences within different valve positions (i.e., $\vartheta_P(t)$), providing various heating powers, the occupancy-rate at time t (i.e., $O_{CC}(t)$), the actual and m-past measures of outdoor temperature (i.e., $T_{Out}(t), T_{Out}(t-1), \ldots$ and $T_{Out}(t-m)$), and the actual and n-past values of indoor temperature (i.e., $T_{In}(t), T_{In}(t-1), \ldots$ and $T_{In}(t-n)$).

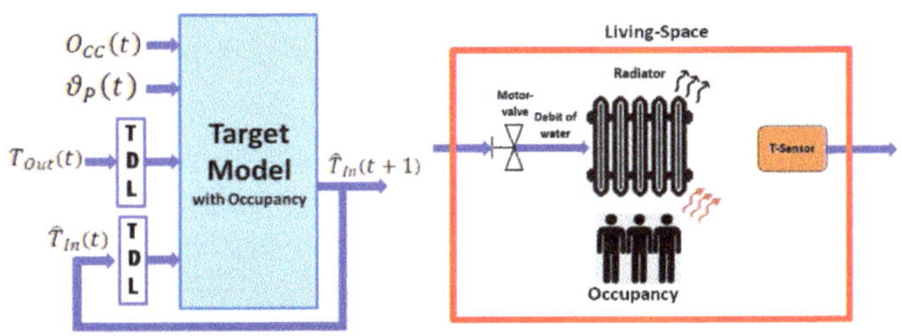

Figure 2. Proposed identification architecture of the target heating model (left-side) and general diagram of the modeled living space including the influence of its occupancy (right-side).

The influence of occupancy is modeled through the following hypothesis: *occupancy of the living space by residents increases the effective overall heating power*. The right-side picture of Figure 2 gives the general diagram of the modeled living-space within the aforementioned hypothesis. In other words, we assume that occupation of the living space by residents (i.e., bodies providing additional heating sources) is equivalent to an incensement of the heating device's nominal power. To determine the above-mentioned equivalent nominal power, we exploit the usual policy of setting the adequate heater's nominal power versus the living space's characteristics. In fact, building designers determine the adequate heater's nominal power on the basis of the volume of the concerned living space (i.e., the living space that is supposed to be heated by the heating device) by keeping constant a parameter called "*Heating Ratio*" (denoted by HR_0, expressed in W/m^3), defined by Equation (3), where P_{No} denotes the heating device's nominal power (which depends on the technological and structural features of the heating device) and V_{LS} denotes the volume of the living space (room, and so on). The appropriate value of HR_0 is determined versus construction norms (materials, processes, and so on used for constructing the concerned building) and urban, social, and environmental regulations (imposed by legal authorities).

$$HR_0 = \frac{P_{No}}{V_{LS}} \tag{3}$$

Taking into account the hypothesis related to the occupancy's influence, Equation (3) may be extended in terms of Equation (4) taking into account the occupancy's influence. In this equation, $HR(N)$ denotes the "*Heating Ratio*" taking into account the occupancy, $P_{OCC}(N)$ states for additional heating power provided by the living space's occupancy (with $P_{OCC}(N = 0) = 0$), and V_{LS} denotes the volume of the living space (room, and so on).

$$HR(N) = HR_0 + \frac{P_{OCC}(N)}{V_{LS}} \tag{4}$$

One can notice that $HR(N = 0) = HR_0$ corresponds to the *Heating Ratio* of the same living space when it is empty. We determine $P_{OCC}(N)$ through a fuzzy-logic-based concept by considering, on the one hand, three categories (fuzzy intervals) of living spaces (i.e., three fuzzy intervals of living space's volume), and on the other hand, by considering five categories of occupancy-rate (denoted by O_{CC}, expressed in %). Namely, the three categories for living space's volume are as follows: "Large", "Medium", and "Small". The five categories of occupancy-rate are as follows: "Empty" (corresponding to $O_{CC} = 0\%$), "Small occupation" (i.e., $O_{CC} = 25\%$), "Medium occupation" (i.e., $O_{CC} = 50\%$), "High occupation" (i.e., $O_{CC} = 75\%$), and "Full" (i.e., $O_{CC} = 100\%$). Within this formulation, Equation (4) may be written in terms of Equation (5), where V_{LS}^F denotes the "*fuzzy value*" of V_{LS}.

$$HR(O_{CC}) = HR_0 + \frac{P_{OCC}(O_{CC})}{V_{LS}^F} \tag{5}$$

The main advantage of such a formulation is that the above-mentioned intervals may be quantified by the use of data-driven statistical clustering methods, reflecting the reality of the concerned building's usage (occupancy). Thus, the equivalent nominal power of a given living space occupied by N bodies may be estimated as $HR(O_{CC}) \times V_{LS}$.

Flooding back to the hypothesis we made related to the effect of the living space's occupancy, another way of interpreting the aforementioned hypothesis is to note that the occupancy of the living space by residents will decrease the required time for reaching the target indoor temperature. In other words, the higher the living space's occupancy, the shorter the necessary delay to heat it. In order to quantify this, we introduce what we call "*Heating Slop*" (denoted by $h(t)$ and expressed in °C/s), defined as the derivative of $T_{In}(t)$ versus the time and approximated by Equation (6) within the context of a discrete sampling (i.e., discrete measurement) of the indoor temperature. In Equation (6), $T_{In}(t_k)$

and $T_{In}(t_{k-1})$ stand for consecutive values of indoor temperature (supposed to be provided by the temperature sensor at times t_n and t_{n-1}, respectively) and $\Delta t = t_k - t_{k-1}$.

$$h(t) = \frac{T_{In}(t_k) - T_{In}(t_{k-1})}{\Delta t} \tag{6}$$

It is pertinent to notice that an escalation of the living space's occupancy (introducing additional bodies and thus additional sources of heating) or decrease of occupancy will result in the so-called heating slop's modification: the higher the living space's occupancy, the stronger the heating slop. Within the general standpoint, and as formulated by Equation (6), $h(t)$ is time-dependent, and thus may vary along with time. However, because of the fact that the heating of buildings abides by slow dynamics, often $h(t)$ remains constant (with regard to the time), albeit its value would vary along with the valve-position (i.e., with $\vartheta_P(t)$) that controls the heating device's actual power. On the basis of the aforementioned points, actually, the predicted indoor temperature (i.e., $\hat{T}_{In}(t_{k+1})$) may be computed from Equation (7), where $\hat{h}(t) = f(\vartheta_P(t)\ ,\ O_{CC})$ denotes the identified (estimated) value of $h(t)$.

$$\hat{T}_{In}(t_{k+1}) = \hat{h}(t)\ (t_{k+1} - t_k) \tag{7}$$

3. Implementation of the Proposed Living-Spaces' Dynamic Heating Model

As has been mentioned, a real five-floor building located at Senart Campus of University Paris-Est Créteil (UPEC) served as an experimental platform for the evaluation and validation of the proposed model. The concerned building (namely Building A of the campus) is a fully automated building hosting the Electrical Engineering and Industrial Informatics Department of Senart-Fontainebleau Institute of Technology of UPEC. The building (i.e., system to be identified) is heated by a conventional central heater supplying radiators (i.e., heating devices) located in various living spaces (namely, office rooms, classrooms, practical rooms, and so on) of the building. The central heater is common to three buildings of the campus, and thus the control of the local heating devices of the concerned buildings (including Building A) is performed through the local valves of each radiator. The two other buildings are conventional buildings (i.e., not automated) and the sole Building A is automated. In fact, Building A is equipped with numerous sensors and connected devices allowing the recording of data related to environmental information (such as temperatures in each living space and the outdoor temperature) and the operational states of whole installed connected devices (such as radiators' valves). Four different kinds of sensors outfit each living space (including corridors) the entire five floors of this building: "temperature sensors" (TSs), "magnetic sensors" (MSs), "presence detectors" (PDs), and "luminance sensors" (LSs). The main connected devices (actuator) deployed in the aforementioned experimental building are as follows: "motor valves" (MVs), which control radiators supplied by the abovementioned central water-flowed heating system, and connected "lighting elements" (LEs).

Sensors and connected devices concerned by the purpose of the present paper are TSs and MVs. They use "EnOcean" technology; an energy harvesting wireless technology provided by EnOcean [27]. EnOcean-technology-based modules fuse micro-energy converters with ultralow power electronics and reliable wireless communications, allowing to provide self-powered wireless sensors or actuators for building energy management systems as well as for industrial applications. Figure 3 presents the implementation diagram of the concerned building (Building A) heating system.

The connected heating system includes three operational layers:

- Supervision layer (SL): It consists of a PC including TopKapi server supervision software (a supervisory control and data acquisition software), which acts as a supervision agent. It also includes a number of adequate interface agents (software units) concerning the control layer and storage memory [28]. It is relevant to note that, while nowadays micro-controllers are able to handle diverse computational skills, they may still be limited regarding computational needs relating to the context of the presented work. In fact, in our work, we deal with machine learning-based identification, where a number of computational tasks need improved computational ability (especially for the training task). Actually, the effective adaptability to the real-usage context of the system would require updating the models' parameters versus the evolution of effective conditions (i.e., bring up to date the system's "knowledge"). That is why the choice was directed toward integrating a server. Moreover, the target system addresses smart-buildings' context, and thus would deal with a rather large number of living spaces. This reinforces the choice of superior computational ability.

- Control layer (CL): This layer contains the programmable logic controller (PLC) and EnOcean modules (pilots and interfaces) necessary to conduct the related sensors and devices composing the physical layer [29]. The concerned PLC is a "WAGO-I/O-SYSTEM" belonging to the family of ETHERNET programmable Fieldbus controllers distributed by WAGO company [30]. It supports both MODBUS/TCP and a wide variety of standards ETHERNET/IP protocols in order to integrate easily into various IT environments.
- Physical layer (PL): It consists of the aforesaid sensors and actuators devices.

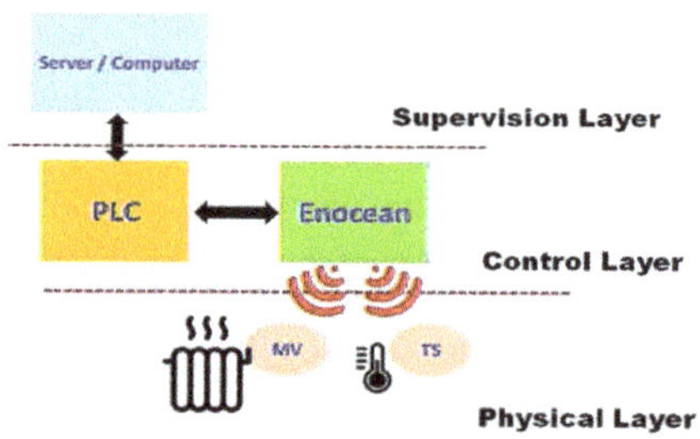

Figure 3. Implementation diagram of the heating system equipping Building A of Senart Campus of University Paris-Est Créteil (UPEC). PLC, programmable logic controller; MV, motor valve; TS, temperature sensor.

Composing the heating control chain of the SBEMS of the aforementioned fully automated experimental building, the CL and PL are replicated for each floor, making possible the set up collecting data characterizing the heating state of each living space of the building and controlling valve position of each heating device (radiator) in the building through the five PLCs (one for each floor). The proposed identification approach was implemented in the SBEMS of the aforementioned fully automated experimental building.

For evaluation of the proposed identification strategy and the issued model, two experimental assessments were considered. The first one appraises the obtained model's "one-step prediction" (OSP) accuracy and the second one sizes up the ability of the issued model on "multi-step prediction" (MSP). The purpose of OSP aims to predict the living space's immediate upcoming indoor temperature from its previous history. Therefore, open-loop as well as closed-loop architectures could be used. Meanwhile, the objective in MSP relates to the prediction of several successive future steps of the concerned living space's indoor temperatures, and thus the open-loop architecture remains no more pertinent.

4. Experimentation and Results

4.1. Experimental Protocol's Description

Both of the two aforementioned evaluations are performed in keeping with the same experimental protocol. This protocol considers a living space of Building A belonging to the category of "middle-size" working spaces of this building (i.e., V_{LS}^{F} = "*Medium*") able to soak up 28 residents (individuals). The considered living space is equipped with a 3 kW heating device (namely a 3160 W radiator supplied by the central heater), responding to the construction and legal norms applicable to this category of working spaces. The collected data are remote values (time history) of outdoor temperature

$T_{Out}(t)$, radiator's valve position $\vartheta_P(t)$, and indoor temperature $T_{In}(t)$. Two sets of experimental data were collected. The data sampling period is one minute, meaning that the value of each considered parameter is collected periodically every 60 s. Figure 4 depicts the experimental conditions showing the valve's position and living space's temperature evolution, respectively.

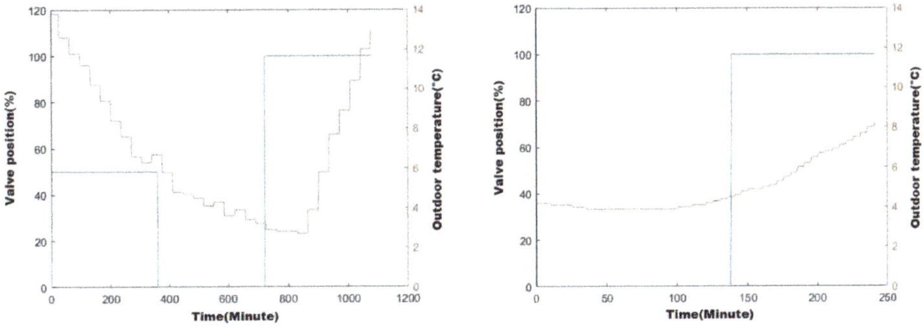

Figure 4. Experimental data collection conditions comply with the empty working space (left-side diagram) and fully occupied working space (right-side diagram), respectively, showing the valve position and outdoor temperature–time history.

The first one, which we tag as "Empty-working-space" (EWS), includes the time history of the aforementioned parameters' values during 18 h (i.e., 1080 min) when the considered working space is empty (i.e., $O_{CC}(t) = 0\%$). The left-side diagram of Figure 4 depicts the experimental conditions, showing the local heating device's operational conduct and the outdoor temperature's evolution during the data collection sequence. As is visible from this diagram, the working space was heated over six hours by the radiator operating with 50% of its nominal power (i.e.,

$\vartheta_P(t) = 50\%$ during the first six hours). Then, the heating was stopped during the six next hours (i.e., $\vartheta_P(t) = 0\%$ from $t = 360'$ to $t = 720'$). Finally, during the last period of six hours, the work space was heated by the radiator developing its maximum nominal power (i.e., $\vartheta_P(t) = 100\%$ from $t = 720'$ to $t = 1080'$).

The second one, which we tag as "Fully-occupied-working-space" (FWS), includes the time history of the aforementioned parameters' values when the same working space is occupied by 28 individuals (i.e., $O_{CC}(t) = 100\%$) during 100 min of a period of four hours. The right-side diagram of Figure 4 depicts the experimental conditions related to this second set of collected data, showing the local heating device's operational conduct and the outdoor temperature's evolution during the data collection sequence. As is visible from this diagram, the working space is not heated during the first 140 min, assuming that the working space is empty and thus does not need to be heated. Then, while being fully occupied (i.e., $O_{CC}(t) = 100\%$) during next period of 100 min, it is heated by the radiator developing its maximum nominal power (i.e., $\vartheta_P(t) = 100\%$ from $t = 140'$ to $t = 240'$).

A part of the collected datasets serves for training of the proposed ANN-based NARX identifier and a ratio of the collected data is used as testing data. The parameters of ANN for all constructed models are as follows:

- Identification of empty working space was performed using ANN including one hidden layer with a size of 5 (number of neurons in the hidden layer). The number of neurons in the hidden layer was set empirically. Related to the training and validation, 85% of data was utilized for training and 15% for testing.
- Identification of fully occupied working space was performed using ANN including one hidden layer with a size of 10 (number of neurons in the hidden layer). The number of neurons in the hidden layer was set empirically. Related to the training and validation, 90% of data was utilized for training and 10% for testing.

The training and testing operations were repeated 10 times for each collected dataset. The evaluation of the obtained model's accuracy was done based on mean squared error (MSE) and mean absolute error (MAE) criteria. It is illustrated by Equations (8) and (9), respectively, where N is number of samples, y_i is the effectively recorded data, and $\hat{y}_i$ represents the estimated (predicted) value.

$$MSE = \frac{1}{N}\sum_{i=1}^{N}(y_i - \hat{y}_i)^2 \quad (8)$$

$$MAE = \frac{1}{N}\sum_{i=1}^{N}\left|y_i - \hat{y}_i\right| \quad (9)$$

It is relatable to note that the first trial concerned the experimental settlement of the number of time-delayed data samples to be considered in the prediction task. The concerned data correspond to the three-hour sequence of the EWS when the considered working space is heated by the radiator developing its nominal power, leading to a swell of 2.7 °C of the indoor temperature (i.e., data recorded between $t = 720'$ and $t = 900'$). A part of extracted sequence was used for identification and the rest of the sequence's data served for testing the forecasting ability of the issued model. The amount of data related to the machine learning task links the representativeness of the collected data regarding the considered environmental and human (i.e., dwellers) factors involved in the constructed models (i.e., influencing the system's conduct). Typical delays of living spaces' heating (i.e., temperature variation) reflecting a representative sequence of the buildings operational behavior are between two hours (for a fully occupied living space) and six hours (for an empty living space). Taking into account the implementation technologies and the precision of the deployed sensors (i.e., 0.1 °C for temperature sensors and 1% for radiators' motor valves), the sampling period (data acquisition every minute), and the involved building's heating dynamics, this leads to to a sufficient amount of data representative of the system's conduct. Besides the above-mentioned, the proposed system's implementation architecture allows a versatile collection of complementary data at any time or in a continuous way.

Twenty models were constructed, differing with regard to the number of the considered time-delayed data samples involved in the prediction task. Figure 5 depicts the obtained results representing the minimum and maximum value of MSE versus the number of the considered time-delayed data samples involved in the prediction task. The above-stated number of models has been aspired by technical features of the deployed implementation technology's features. Actually, the sampled data are transmitted by the deployed module every 18 min with the already mentioned sampling period of 60 s. By taking this fact into account, we aimed to study the plausible influence on the forecasting accuracy of considered time-delayed data (from 1 to 20) related to the involved parameters (i.e., indoor temperature, valve position). On the other hand, each model has been trained and tested 10 times, allowing a quantitative (i.e., statistics) evaluation of the aforementioned possible influence.

If MSE_{min} remains within the interval $[0.12\ ,\ 0.26]$, the lowest obtained MSE_{max} values are obtained with $n = 4$ and $n = 20$. Taking account of the implementation's computational constraints, $n = 4$ (corresponding to $MSE_{min} \cong 0.12$ and $MSE_{max} < 1.5$) appears to offer a suitable compromise. According to the obtained results (shown in Figure 5), the experimental evaluation was performed using n = 4, stressing our choice toward a lower computational complexity.

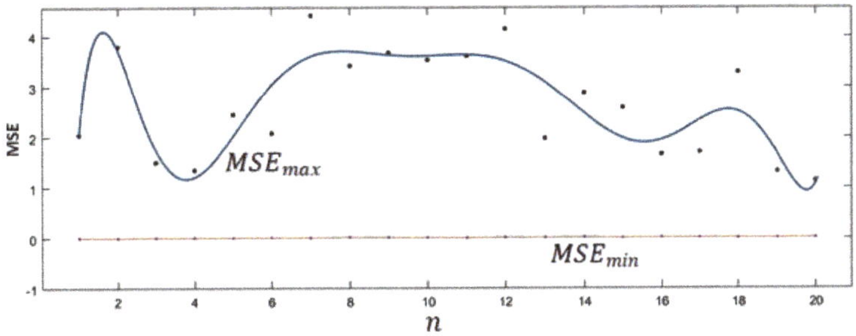

Figure 5. Minimum and maximum mean squared error (MSE) (MSE_{min} and MSE_{max}) versus the number of considered time-delayed data samples involved in the prediction task (n).

4.2. Experimental Results

The two constructed datasets have served for evaluating OSP as well as MSP models of the considered working space's heating conduct. Following the experimental protocol described in the previous section, the data-driven learning-based issued model was pointed out using data partly from EWS and partly from FWS. Two operational scenarios corroborating experimental recorded data conditions (i.e., those depicted by Figure 4) were considered:

1. The first case study, focusing OSP accuracy, considers two painless operational situations:

    a The first situation assumes that the considered working space is empty (i.e., $O_{CC}(t) = 0\%$), the outdoor temperature is up to $T_{Out}(t) = 10$ °C, and the indoor temperature at the beginning of the heating sequence is 20.86 °C (i.e., $T_{In}(t = 0) = 20.86$ °C). Starting under the above-mentioned conditions, the considered empty working space is supposed to be heated during two hours (i.e., during 120′) by the radiator developing its nominal heating power (i.e., $\vartheta_P(t) = 100\%$).

    b The second situation assumes that the considered working space is fully occupied by 28 residents (i.e., $O_{CC}(t) = 100\%$), the outdoor temperature is up to $T_{Out}(t) = 5$ °C, and the indoor temperature at the beginning of the heating sequence is 21.80 °C (i.e., $T_{In}(t = 0) = 20.80$ °C). Starting under the above-mentioned conditions, the considered occupied working space is supposed to be heated during 350′ by the radiator developing its nominal heating power (i.e., $\vartheta_P(t) = 100\%$).

2. The second case study focuses on MSP accuracy evaluation, considering two more tricky situations:

    c Assuming that the considered working space is empty (i.e., $O_{CC}(t) = 0\%$), the first situation of this second case-study presumes that the working space is heated in accordance with the left-side diagram of Figure 4. In other words, it supposes that the radiator heating the considered living space heats it during six hours developing 50% of its nominal power (i.e., $\vartheta_P(t) = 50\%$ for $t \in [0', 360']$), and then after a six-hour halt (i.e., $\vartheta_P(t) = 0\%$ for $t \in [361', 720']$), it reheats this same working space during an additional six hours developing its whole nominal power (i.e., $\vartheta_P(t) = 100\%$ for $t \in [721', 1080']$). The outdoor temperature (i.e., $T_{Out}(t)$) is supposed to vary during those 18 h within the interval [2 °C , 14 °C], also in line with the left-side diagram of Figure 4. The indoor temperature at the beginning of the heating sequence is 19.60 °C (i.e., $T_{In}(t = 0) = 19.60$ °C).

    d Presuming that the considered working space is empty at the beginning (i.e., $O_{CC}(t) = 0\%$ at t = 0), the second situation of this second case-study assumes that the concerned working space becomes fully occupied during 100 min and is reheated in accordance with the right-side diagram of Figure 4. In other words, it supposes that the radiator is off during the first 140 min when the living space is empty (i.e., $\vartheta_P(t) = 0\%$ and $O_{CC}(t) = 0\%$ for $t \in [0', 140']$), and it heats the considered living space during 100 min when the room is fully occupied, developing its nominal power (i.e., $\vartheta_P(t) = 100\%$ and $O_{CC}(t) = 100\%$ for $t \in [141', 240']$). The outdoor temperature (i.e., $T_{Out}(t)$) is also supposed to vary within the interval [4 °C, 8 °C] in line with the right-side diagram of Figure 4. The indoor temperature at the beginning of the heating sequence is 17.25 °C (i.e., $T_{In}(t = 0) = 17.25$ °C).

Figures 6 and 7 show the obtained results related to the two above-mentioned case-studies. Concerning the first case-study, the left-side diagram of Figure 6 plots the estimated (i.e., model-based OSP) indoor temperature and measured (i.e., real) indoor temperature of the considered living space when it is empty. The right-side diagram of this same figure gives model-based predicted and real indoor temperature's values when the considered living space is fully occupied. Linking the second case-study, the left-side diagram of Figure 7 gives the estimated (i.e., model-based MSP) indoor temperature and measured (i.e., real) indoor temperature of the considered living space when it is empty. The right-side diagram of this same figure depicts the model-based, multi-step prediction of indoor temperature values and the measured temperature when the living space is fully occupied during 100 min.

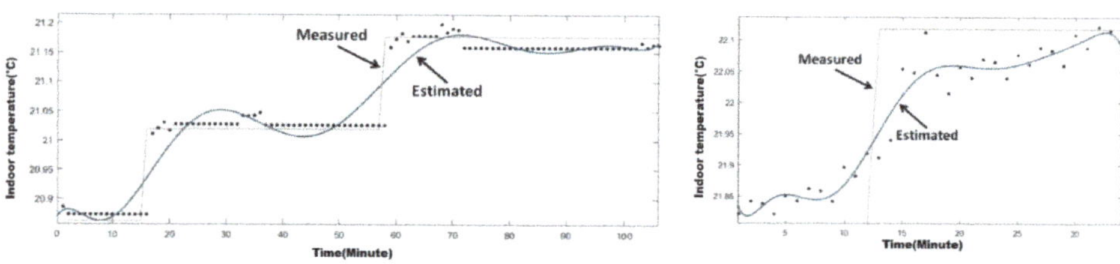

Figure 6. Model-based, one-step predicted indoor temperature time history when the living space is empty (left-side) and when it is fully occupied (right-side).

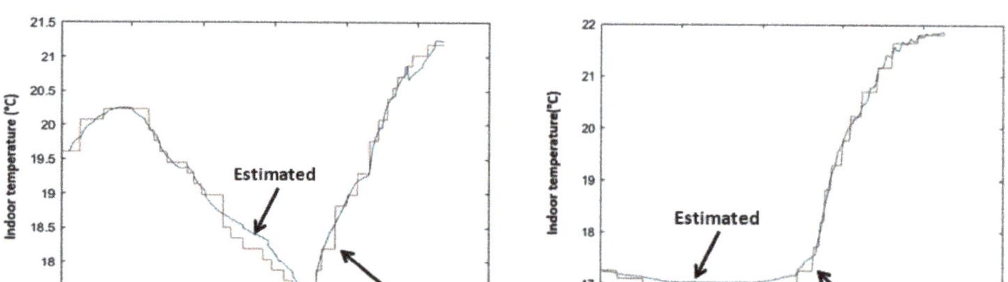

Figure 7. Model-based, multi-step predicted indoor temperature time history when the living space is empty (left-side) and when it is fully occupied (right-side).

Table 1 summarizes the overall accuracy of OSP and MSP models of the considered working space's heating conduct. As expected, the OSP model of the aforementioned heated living space forecasts the upcoming value of the indoor temperature with less than 0.2 °C blunder compared with the measured value. Actually, the achieved high prediction accuracy is because of the fact that, in the OSP model, the prediction is performed using the four effectively-measured past values of the indoor temperature, and thus representing the effective time-history of indoor temperature's evolution. However, anchored in an open-loop NARX scheme, the main shortage of this model would appear when a longer-term forecasting of indoor temperature is needed.

Table 1. Estimation's maximum and minimum mean squared error (MSE) and mean absolute error (MAE) with four delayed samples considered for each involved parameter (i.e., $n = 4$). OSP, one-step prediction; MSP, multi-step prediction.

Tests / Errors	OSP	MSP (Test Data)	MSP (Whole the Data)
MSE_{min}	0.0006	0.021	0.020
MAE_{min}	0.01	0.12	0.11
MSE_{max}	0.03	2.45	0.25
MAE_{max}	0.15	1.23	0.44

The MSP model of the aforementioned heated living space forecasts the upcoming value of the indoor temperature with lower accuracy compared with the OSP model; an average error up to 0.4 °C, while attaining, for some long-term predicted values of indoor temperature, an error exceeding 1.23 °C. The finer analysis of indoor temperature's forecasting, supported by the results depicted in Figure 8, provide the incentive of the observed gap. The left-side diagram of Figure 8 reports the estimation's absolute error for 35 consecutive estimated indoor temperatures (i.e., prediction of the 35 upcoming values of indoor temperature) of the modeled heated working space. The right-side diagram of Figure 8 plots the forecasting error of the so-called "heating slop" (i.e., $h(t)$), defined and introduced in Section 2. In fact, as visible from those diagrams, if both estimation errors (i.e., the estimation error relative to $T_{In}(t)$ and the forecasting error related to $h(t)$ values' estimation) remain close to zero for short-term (5 values) and middle-term (15 values) forecasted indoor temperature values, both of them admit a continuously increasing evolution for long-term predicted values, especially those surpassing the next thirty-minute predicted period. Actually, within such a longer-term prediction requirement, the generalization of MLP neural net seems to reach its limitation regarding the learning dataset.

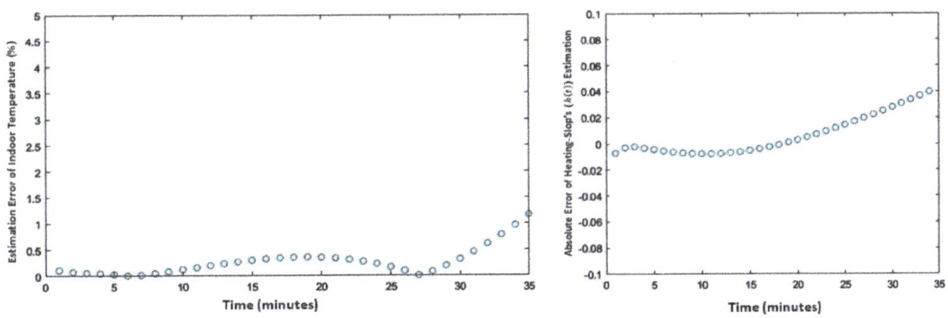

Figure 8. Indoor temperature estimation's absolute error (left-side) and heating slop's forecasting error (right-side).

Figure 9 depicts the heating slop (i.e., $h(t)$) when the considered living space is heated by its radiator in two different situations: the blue curve corresponds to $h(t)$ when the living space is empty and the other curve relates to the situation when the same heated space is fully-occupied. It is pertinent to notice that the average estimated heating slop for the empty living space is $\overline{h(t)} = 0.015 \frac{°C}{minute}$, while, for the same fully occupied living space, it is more than four times stronger ($\overline{h(t)} = 0.042 \frac{°C}{minute}$), showing up the occupancy's impact on the considered living space's heating dynamics. In fact, the indoor temperature of the considered living space heated with a same radiator will increase up to 0.5 °C in 35 min when it is empty, while it will increase up to 1.5 °C when the living space is fully occupied.

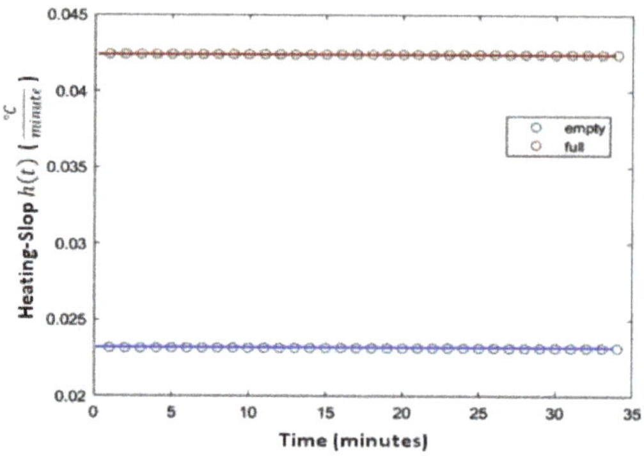

Figure 9. Heating slop when the living space is empty and when it is fully occupied.

Finally, Figure 10 shows $\frac{P_{OCC}(O_{CC})}{V_{LS}^{F}}$ (introduced in Equation (5) of the second section) versus the occupancy rate for the three considered categories of living spaces of Building A, where $O_{CC} = 100\%$ corresponds to the occupancy of each considered living space category's by up to 28 individuals. As an example, according to this diagram, 28 individuals occupying a medium living space of this building, equipped with a 3 kW radiator, would correspond to an additional heating power of 1170 W.

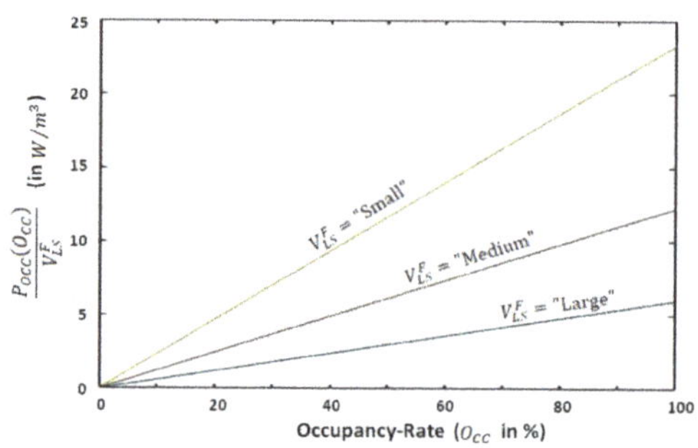

Figure 10. $\frac{P_{OCC}(O_{CC})}{V_{LS}^{F}}$ (introduced in Equation (5) of the second section) versus the occupancy rate for the three considered categories of living spaces of Building A.

5. Conclusions

Considering that, alongside the living space's intrinsic structural features, the occupancy of the living space (by users or residents) may affect the model of heating dynamics of the concerned living space, we have investigated the design, implementation, and validation of a data-driven machine learning-based identifier supplied by the time-series prediction paradigm's formalism. The proposed data-driven machine learning-based identifier was applied for modeling the dynamic heating conduct of a real fully automated five-floor building's living spaces located at Senart Campus of University Paris-Est Créteil, taking into account their occupancy (by users of this public building). The proposed modeling strategy takes advantage, on the one hand, of the time-series' forecasting capacity of the NARX model, and on the other hand, of the multi-layer perceptron's (MLP) learning and generalization skills.

If, as expected, the one-step-prediction (OSP) model, operating on the basis of an open-loop scheme, achieved high prediction accuracy in forecasting of the upcoming value of the indoor temperature (i.e., less than 0.2 °C blunder comparing to the measured value), anchored in an open-loop NARX scheme, its main shortage appears when a longer-term forecasting of indoor temperature is required; especially, when the target model is used for designing an adaptive heating control strategy. Achieving a lower accuracy compared with the OSP model (i.e., an average error up to 0.4 °C and a maximum error of

1.23 °C for long-term prediction), the multi-step-prediction (MSP) model, operating in closed-loop, represents an attractive compromise for longer-term forecasting of the dynamic heating behavior, and thus offers an appealing perspective for designing adaptive heating controllers for SBEMSs.

The achieved results stress several appealing issues related to the denotation of these results as well as the status of the NARX-based forecaster regarding data-driven identification of heating dynamics in real smart-buildings. The first points come across the ability of the proposed approach in the modeling complex thermal conduct of buildings, including the effect of inhabitants' presence on the discrepancy of their heating dynamics. In fact, this is visible through the obtained MSE and MAE values, highlighting a prediction of indoor temperature with a less than 0.2 °C blunder compared with the measured value. This foretells the perspective of effectual usage of the proposed approach for designing data-driven adaptive controllers of buildings' heating behavior versus the context of their usage by potential residents. The second remark relates to the possibility of a standard-technology-based effective implementation of this investigated machine learning-based identifier in authentic smart-buildings, taking advantage of the robustness of those standard (and market available) technologies and avoiding the complexity and cost of designing specific implementation policies. Finally, the last mention goes to the accuracy of the achieved predictions related to well short-term (i.e., one-step) as well as long-term (i.e., closed-loop) forecasters. Another attractive feature, arising from the theoretical foundation of the proposed approach, relates to the comprehensive interpretation of the living space's occupancy effect, with a quantitative appreciation of its influence on the smart-building's heating conduct's deviation.

Author Contributions: Conceptualization, R.S.B., K.M. and A.C.; Data curation, R.S.B.; Investigation, R.S.B.; Methodology, R.S.B.; Resources, R.S.B., A.C., V.A. and L.H.; Software, R.S.B.; Supervision, K.M.; Validation, R.S.B.; Visualization, R.S.B.; Writing–original draft, R.S.B.; Writing–review & editing, K.M. All authors have read and agreed to the published version of the manuscript.

Funding: This research received no external funding.

Conflicts of Interest: The authors declare no conflict of interest.

References

1. U.S. EIA. Energy Use in Commercial Buildings-Energy Explained, Your Guide To Understanding Energy—Energy Information Administration. Washington, DC, USA, 2017. Available online: https://www.eia.gov/energyexplained/index.php?page=us_energy_commercial (accessed on 4 September 2019).
2. Space Heating and Water Heating (EIA). Available online: https://www.eia.gov/todayinenergy/detail.php?id=37433 (accessed on 4 September 2019).
3. Alhamoud, A.; Ruettiger, F.; Reinhardt, A.; Englert, F.; Burgstahler, D.; Böhnstedt, D.; Gottron, C.; Steinmetz, R. An Intelligent System for Energy Saving in Smart Home. In Proceedings of the 39th Annual IEEE Conference on Local Computer Networks Workshops, Edmonton, AB, Canada, 8–11 September 2014; pp. 685–692.
4. James, D.H. *The Radiation of Heat from the Human Body New York Hospital*; PubMed Central (PMC): New York, NY, USA, 1934; pp. 615–620.
5. Fang, J.S.; Hao, Q.; Brady, D.J.; Shankar, M.; Guenther, B.D.; Pitsianis, N.P.; Hsu, K.Y. Path-Dependent Human Identification Using a Pyroelectric Infrared Sensor and Fresnel Lens Arrays. *Opt. Express* **2006**, *14*, 609. [CrossRef] [PubMed]
6. Johnson, C. *Mathematical Physics of BlackBody Radiation*; Icarus iDucation: Odin Teatret, Denmark, 2012.
7. Zungeru, A.M.; Mangwala, M.; Chuma, J.; Gaebolae, B.; Basutli, B. Design and Simulation of an Automatic Room Heater Control System. *Heliyon* **2018**, *4*, e00655. [CrossRef] [PubMed]
8. Aaurav, K.; Ch andan, V. Gray-Box Approach for Thermal Modelling of Buildings for Applications in District Heating and Cooling Networks. In Proceedings of the 8th International Conference on Future Energy Systems, Hong Kong, China, 16–19 May 2017; pp. 347–352.
9. Purdon, S.; Jurdak, B.K.R.; Challen, G. Model-free HVAC control using occupant feedback. In Proceedings of the 38th Annual IEEE Conference on Local Computer Networks, Sydney, Australia, 21–24 October 2013.
10. Asgari, H.; Chen, X.Q.; Menhaj, M.B.; Sainudiin, R. ANN-Based System Identification, Modelling and Control of Gas Turbines—A Review. *Adv. Mater. Res.* **2013**, *622*, 611–617. [CrossRef]

11. Wiener, N. Non-Linear Problems in Random Theory. In *Physics Today*; American Institute of Physics: College Park, MD, USA, 1959; Volume 12, p. 131.
12. Schetzen, M. *The Volterra And Wiener Theories of Nonlinear Systems*; Wiley: New York, NY, USA, 1980.
13. Takagi, T.; Sugeno, M. Fuzzy Identification of Systems and Its Application to Modeling and Control. *IEEE Trans. SMC* **1985**, *5*, 116–132. [CrossRef]
14. Boussaada, Z.; Curea, O.; Remaci, A.; Camblong, H.; Mrabet Bellaaj, N. A Nonlinear Autoregressive Exogenous (NARX) Neural Network Model for the Prediction of the Daily Direct Solar Radiation. *Energies* **2018**, *11*, 620. [CrossRef]
15. Brus, L. Nonlinear Identification of a Solar Heating System. In Proceedings of the IEEE Conference on Control Applications (CCA 2005), Toronto, ON, Canada, 28–31 August 2005; pp. 1491–1497.
16. Rabbani, M.J.; Hussain, K.; Khan, A.U.R.; Ali, A. Model Identification and Validation for a Heating System Using Matlab System Identification Toolbox. *IOP Conf. Ser. Mater. Sci. Eng.* **2013**, *51*, 012022. [CrossRef]
17. Lahane, J.S.; Meera, A.K. System Identification and Controller Design for Boiler and Heat Exchanger Set-Up. *Int. J. Adv. Res. Electr. Electron. Instrum. Eng.* **2014**, *3*, 10708–10714. [CrossRef]
18. Knudsen, M.D.; Rasmus, E.; Hedegaard, T.H. Pedersen, and Steffen Petersen. System Identification of Thermal Building Models for Demand Response—A Practical Approach. *Energy Procedia* **2017**, *122*, 937–942. [CrossRef]
19. Building Controls Virtual Test Bed. Available online: https://simulationresearch.lbl.gov/projects/building-controls-virtual-test-bed (accessed on 4 September 2019).
20. Open Modelica. Available online: https://www.openmodelica.org/ (accessed on 4 September 2019).
21. Van Overschee, P.; De Moor, B. N4SID: Subspace Algorithms for the Identification of Combined Deterministic-Stochastic Systems. *Automatica* **1994**, *30*, 75–93. [CrossRef]
22. Gao, Y.; Er, M.J. NARMAX time series model prediction: Feed-forward and recurrent fuzzy neural network approaches. *Fuzzy Sets Sytems* **2005**, *150*, 331–350. [CrossRef]
23. Johansen, T.A.; Foss, A.B. Constructing NARMAX using ARMAX. *Int. J. Control* **1993**, *58*, 1125–1153. [CrossRef]
24. Billings, S.A. *Nonlinear System Identification: NARMAX Methods in the Time, Frequency, and Spatio-Temporal Domains*; John Wiley & Sons: Chichester, UK, 2013.
25. Rumelhart, D.E.; Hinton, G.E.; Williams, R.J. *Learning Internal Representations by Error Propagation*; Parallel Distributed Processing: Explorations in the Microstructure of Cognition; MIT Press: Cambridge, MA, USA, 1986; Volume 1.
26. Hastie, T.; Tibshirani, R.; Friedman, J. *The Elements of Statistical Learning: Data Mining, Inference, and Prediction*; Springer: New York, NY, USA, 2009; ISBN 978-0-387-21606-5.
27. EnOcean—The World of Energy Harvesting Wireless Technology, January 2016. Available online: https://www.enocean.com/fileadmin/redaktion/pdf/white_paper/WhitePaper_Getting_Started_With_EnOcean_v1.0.pdf (accessed on 4 September 2019).
28. Sadeghian Broujeny, R.; Madani, K.; Chebira, A.; Hurtard, L. A multi-layer system for smart-buildings' functional and energy-efficiency awareness: Implementation on a real five-floors building. In Proceedings of the 2017 IEEE 8th International Conference on Awareness Science and Technology (iCAST), Taichung, Taiwan, 8–10 November 2017; pp. 87–92. [CrossRef]
29. Sadeghian Broujeny, R.; Madani, K.; Chebira, A.; Hurtard, L. A Machine-Learning Based Approach for Data-Driven Identification of Heating Dynamics of Buildings' Living-Spaces. In Proceedings of the 10th International Conference on Intelligent Data Acquisition and Advanced Computing Systems (IEEE/IDAACS 2019), Metz, France, 18–21 September 2019; Volume 1, pp. 197–202.
30. WAGO Kontakttechnik GmbH and Co. KG, WAGO-I/O-SYSTEM 750 Manueal, 2016. Available online: http://www.safetycontrol.ind.br/imgs/downloads/manual-750-406-pdf-5b211d6fe9baa.pdf (accessed on 4 September 2019).

sensors

MDPI

Article

Improved Pixel-Level Pavement-Defect Segmentation Using a Deep Autoencoder

Rytis Augustauskas * and Arūnas Lipnickas

Department of Automation, Kaunas University of Technology, 51367 Kaunas, Lithuania; arunas.lipnickas@ktu.lt
* Correspondence: rytis.augustauskas@ktu.lt; Tel.: +370-61916583

Received: 31 March 2020; Accepted: 28 April 2020; Published: 30 April 2020

check for updates

Abstract: Convolutional neural networks perform impressively in complicated computer-vision image-segmentation tasks. Vision-based systems surpass humans in speed and accuracy in quality inspection tasks. Moreover, the maintenance of big infrastructures, such as roads, bridges, or buildings, is tedious and time-demanding work. In this research, we addressed pavement-quality evaluation by pixelwise defect segmentation using a U-Net deep autoencoder. Additionally, to the original neural network architecture, we utilized residual connections, atrous spatial pyramid pooling with parallel and "Waterfall" connections, and attention gates to perform better defect extraction. The proposed neural network configurations showed a segmentation performance improvement over U-Net with no significant computational overhead. Statistical and visual performance evaluation was taken into consideration for the model comparison. Experiments were conducted on CrackForest, Crack500, GAPs384, and mixed datasets.

Keywords: CNN (Convolutional neural networks); deep learning; pavement defects; residual connection; attention gate; atrous spatial pyramid pooling

1. Introduction

Continuous pavement monitoring can be an extremely tedious task for humans but easy for automated computer-vision (CV)-based systems. As stated in [1], transportation infrastructure is the backbone of a nation's economy that should be systematically improved. Despite the advantages of modern CV-based systems, real-life applications still meet plenty of challenges. These are usually cases when human experts can identify a surface's three-dimensional defect from first glance, but classical image analysis techniques still fall behind. Therefore, researchers are constantly seeking new approaches to address these challenges.

In ImageNet competitions [2], it was proven that deep neural network-based solutions surpass classical CV methods in object detection by engaging many layers of data abstraction. A data-driven deep-learning (DL) approach might take into consideration a wider spectrum of cases appearing in a complicated problem than can be constrained by only fundamental knowledge.

Difficulties of defect identification can be seen in many manufacturing areas; solutions and the newest method applications were presented in [3] with intelligent imaging and analysis techniques applied in various research fields. The authors in [1] reviewed DL network application publications on pavement crack detection since its first appearance in 2016. The importance of the matter can be seen from a recent review paper on pavement defect detection methods [4]. The number of reviewed methods/publications exceeds 100, and half of them are not older than five years. The authors of [4] and [1] showed how important automated visual crack detection for traffic safety is. Unfortunately, the effectiveness of published approaches is often questioned because of the results being demonstrated on publicly unavailable data.

Pavement and concrete have a diversity of surface structures, rubbish, and drawings (lines or figures painted on roads) that hamper the ability to identify true anomalies, namely, cracks. Although a severely damaged surface is fairly easy to spot, defects that start to form are hardly noticeable. There can be found various approaches to pavement defect detection by utilizing different techniques. Older research papers (written five years ago or more) mostly proposed classical image-processing methods to extract road cracks. Researchers in their work use techniques, such as bilevel [5] or histogram-based [6,7] thresholds, to extract decayed pavements' parts. The authors in [8] used Otsu's [9] threshold for pavement extraction and Sobel [10] filters along the X and Y directions for crack edge signification and segmentation. The researchers in [6] were interested in detecting extremely noticeable road damage, namely, potholes. Furthermore, a data-driven approach for road pavement evaluation was described by Cord et al. [11]. In [12], the Adaboost machine learning method was used to train image classifiers by pavement patches on a 128 × 128 pixel size for quality inspection. However, they described a drawback of the method when dirt appeared on the image. By looking at more recent research publications, most investigations rely on deep learning methods rather than handcrafted image feature classifiers. Even with small neural network architectures, impressive results compared with other techniques can be achieved [13]. Carr at al. [14] received good results by using a simple U-Net [15] (encoder–decoder) convolutional neural network for defective area segmentation from the CrackForest dataset [16,17]. Another interesting approach was introduced in [18] by utilizing the previously mentioned U-Net and additional outputs to every decoding part layer, and afterwards joining them to improve segmentation. This proposed architectural design significantly boosted the prediction accuracy. Moreover, in the combined method in [19], the authors used preprocessing for a concrete image and engaged in deep neural network-based classification. According to the classification output, the segmentation technique based on a threshold in a 2-D histogram was applied later. Additionally, fully integrated solutions were found for pavement distress segmentation, such as that described in [20]. A vehicle with a camera attached to the top was used to scan the road. The authors proposed an architecture that was not very deep for pavement patch classification. Later on, the same researchers proposed a solution [21] with an extended dataset and improved architectural decisions for road defect detection. They utilized various depth models inspired by ResNet [22] to find the most suitable decision. Research was extended by a more advanced neural network model. The authors were only using 2-D image data, but other approaches can be found for a complete pavement defect detection system (camera on a vehicle) [23] that use 3-D depth data for road structure decay detection. However, steerable matched filter banks are utilized instead of deep learning techniques. Furthermore, fully autonomous systems that utilize deep learning techniques along with computer-vision methods for tunnel concrete quality inspection were described in [24,25]. As can be seen, the interest to cope with infrastructure and pavement defect detection problems is still relevant, aiming to apply DL techniques for real-time automated analysis.

In this work, we continued our investigation on computer vision-based pavement crack segmentation by utilizing a convolutional neural network. This is an extension of our previous work [26] presented at the IDAACS'2019 conference. The mentioned article mainly focused on the classical U-Net encoder–decoder architecture depth (number of convolutional layers) and convolutional filter size dependency on the model prediction precision and computational time. All previous experiments were conducted on the CrackForest dataset. In this paper, we extended our research by utilizing two additional datasets from the similar crack-type area, GAPs384 [18,20,21] and Crack500 [18]. In this work, we present a performance analysis of networks trained on mixed data performance on individual datasets. Small additional adjustments were also made using pretrained weights on the targeted dataset to improve the result. Moreover, we introduced architectural convolutional neural network solutions as an improvement to our previous work. Statistical and visual evaluations were taken into consideration. Neural network implementation and all rendered results can be found in a GitHub repository [27].

The paper is organized as followed. Section 2 explains the neural network and its architectural solution, and methods utilized for research. In Section 3, we describe the CrackForest, Crack500, and GAPs384 datasets, and how we used them for our research. Equipment for the experiments and measurement parameters used for evaluating the neural network performance are outlined in Section 4. Results are given in Section 5. Conclusions and discussion are written in Section 6.

2. Deep Neural Network Model

For the baseline in this research, we chose the U-Net [15] deep neural network as an autoencoder function to detect pixel-level cracks in images. The architecture consisted of two main parts, contractive (encoder) and expansive (decoder). The model is shown in Figure 1. In addition to the original structure described in [15], we added padding to the convolutional layers to maintain the output dimension equal to the given input image.

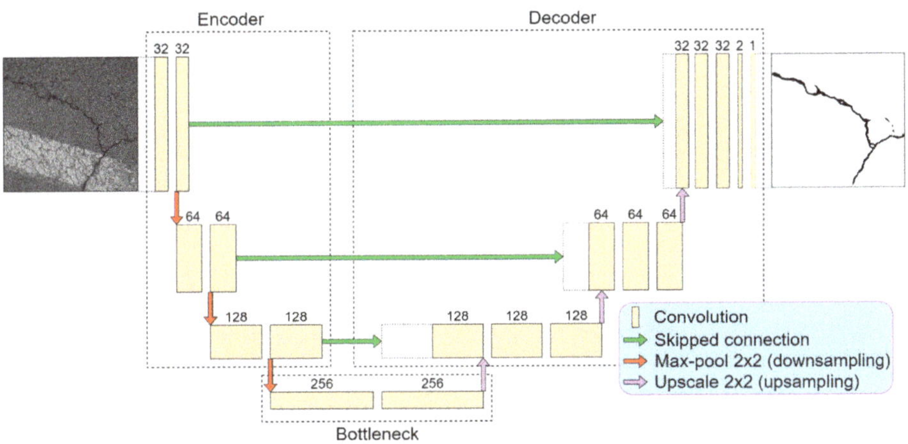

Figure 1. U-Net convolutional neural network structure.

From our previous research [26], we took the best-performing solution for crack detection. It was observed that a deeper convolutional neural network tends to learn features more accurately than smaller architectures. A four-layer (eight convolutional layers in the encoder) neural network outperformed two- and three-layer solutions, although close results in smaller models were achieved by using big feature kernels, such as 7×7 or 9×9. However, big-sized kernel utilization and a small stride significantly slow down the data-processing time. In respect of this, a four-layer neural network was chosen (Figure 1) for this study. All convolutional operations in the encoder part were performed by 3×3 kernels with a one-pixel stride. After every two convolutional operations (until the "bottleneck"), dimensions were reduced twice by a 2×2 max-pooling operation, and the number of features was therefore doubled. The most contracted part is the "bottleneck" that represents a latent space and has the highest number of convolutional kernels. Further, the decoder or reconstruction part starts (Figure 1). With every layer, the tensor width and height dimension were upscaled twice. Afterwards, a 2×2 convolutional operation was performed to adjust and interpret the upscaled data details with the learned parameters. Then, the partly decoded data were concatenated with data from the opposite side (encoder) that transferred higher-level features from the encoder side (see Figures 1 and 2). In all convolutional layers, the rectified linear unit (ReLU) [28] was used as an activation function, and the neural network output (1×1 convolution) had sigmoid activation that output the probability of "how likely it is for a pixel to be a defect" in ranges from 0.0 to 1.0. This corresponds to the range from 0 to 255 in 8-bit grayscale. A higher pixel value meant greater confidence that the pixel belonged to a pavement crack.

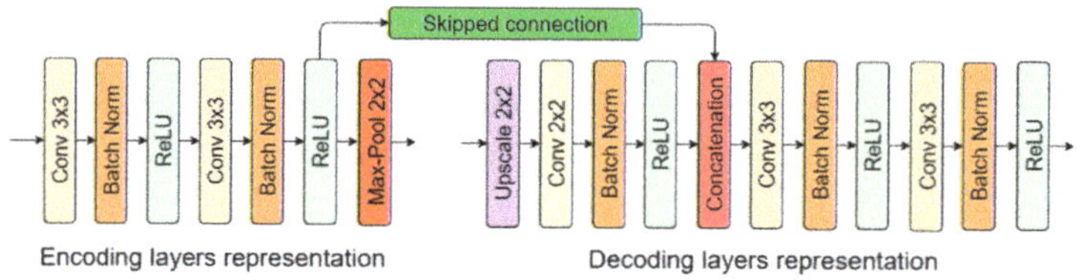

Figure 2. Encoding- and decoding-part layer representation.

Additional to previous research [26], batch normalization [29] was added between the convolutional layer and its activation function (except for the last output layer with sigmoid) to increase the neural network stability. When the mean and variance are calculated using batch normalization for a small minibatch compared to the whole dataset (in this case, 4), it gives noise related to the individual iteration. For this mentioned reason, dropout was removed. Weight decay (L2 normalization) was taken out because batch normalization eliminates its effect [30]. The network-encoding and -decoding-layer representation is shown in Figure 2.

Additionally to the classical U-Net architecture, we applied a few architectural improvements to increase the neural performance. The architecture induced with the residual connections, atrous-spatial pyramid pooling (ASPP), and attention gate (AG) modules is shown in Figure 3. Every modification idea is described briefly in the following subchapters. We conducted experiments with several models: U-Net, U-Net with residual connection, U-Net with residual connection and ASPP module (two types), and U-Net with residual connection, ASPP (2 types), and AG modules. The main aspects of this research were computation and prediction–performance difference investigations.

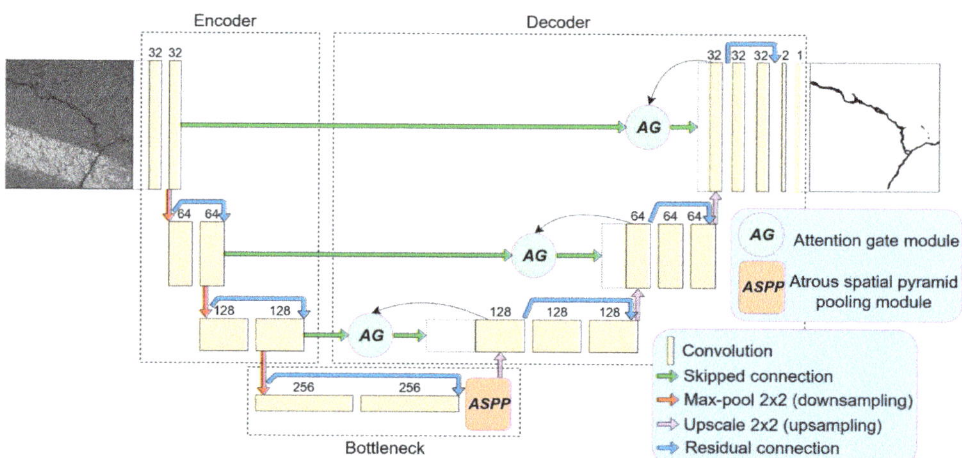

Figure 3. U-Net neural network with few modifications: residual connection, atrous spatial pyramid pooling (ASPP), and attention gate (AG) modules.

2.1. Residual Blocks

Architecture induced with residual connection has been used by multiple researchers [31–34]. It was proven that residual connection helps to fight the vanishing gradient problem, accuracy degradation [35], and improves neural network performance [31,32,34]. Skipped operations also allow undisturbed dataflow through the whole network (Figure 3). In the implementation of the residual connection, we also added 1×1 convolution to adjust the number of features because, in every encoding (downscale) or decoding (upscale), the number changes twice. A residual connection with a double convolutional operation is shown in Figure 4. For the present research, we utilized the original implementation of the residual block proposed in [35].

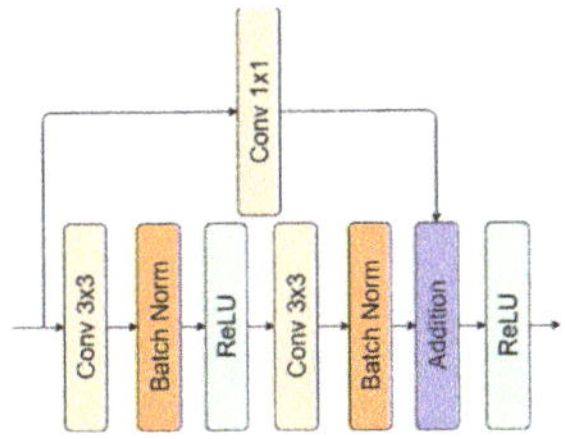

Figure 4. Residual connection representation: 1 × 1 convolutional operation used to adjust number of features.

2.2. Atrous (Dilated) Convolution Blocks

Sequences of dilated convolutions were introduced in [36] as a more capable method to extract semantic information in object segmentation problems. An operation with different dilation rates can take into consideration the multiscale context by utilizing a sequence of convolutions. An example of performing convolutions with different dilation factors is shown in Figure 5.

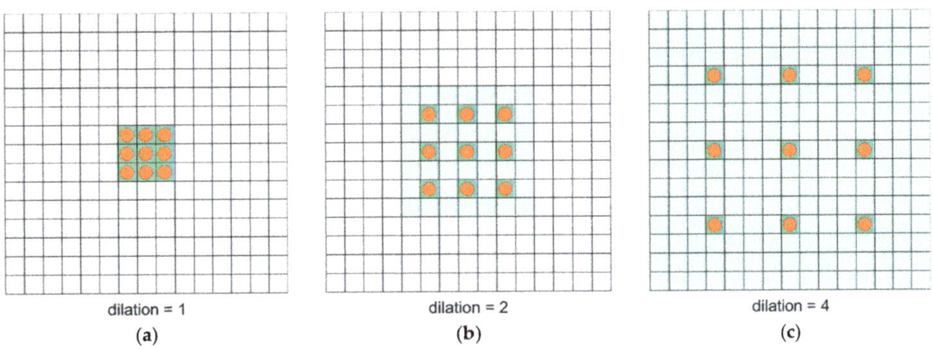

Figure 5. Convolutional 3 × 3 operation with different dilation rates: (**a**)—1, (**b**)—2, (**c**)—4. Orange circles, filter kernel points; green area, contextual image information that might be taken into consideration.

Further studies [37,38] proposed an approach of conducting these operations in parallel. Moreover, in [38], global feature tensor pooling in parallel with a convolutional operation was added to capture global context information with each tensor feature layer, as proposed in ParseNet [39]. The described block (ASPP) is used in state-of-the-art object-detection solutions, such as the DeepLabV3 [38] neural network. Models that were induced with this method showed a performance improvement in satellite images [40–42], medical [43], and general object segmentation tasks [44]. Additionally, to the existing ASSP module structure, the Waterfall connection sequence was introduced in [45], which reused convolutional operations from different parallel convolution, and it outperformed the original (parallel) implementation in object segmentation tasks. Every convolution operation in parallel branches takes the previous convolution result as an input. In this work, the convolution with 1, 2, and 4 dilation rates was used, considering the small pavement crack scale invariance. The picked values were more likely to be intuitive, and the different choice of dilation factors might yield worst or better results, as was shown in experiments conducted in satellite image segmentation [41]. We tested two types of ASPP blocks:

- As proposed in [38], convolutional operations were performed separately in parallel (Figure 6a); and

- As proposed in [45], input from the previous branch (Figure 6b) was reused for convolutional operations.

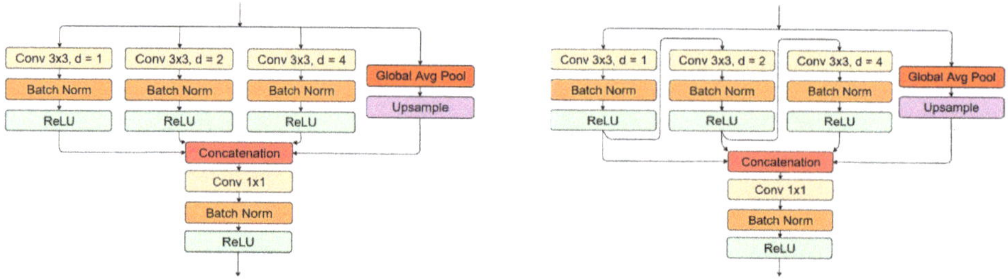

(a) (b)

Figure 6. Atrous spatial pooling block representation: (**a**) ASSP block of convolutions with 1, 2, and 4 dilation rates and global pooling in parallel; (**b**) ASPP block with convolutions with 1, 2, and 4 dilation rates connected as suggested in [45] and global pooling in parallel.

2.3. *Attention Blocks (Attention Gates)*

Attention maps were originally proposed in [46] as a technique to improve image classification. Attention modules highlight relevant and suppress misleading information, such as the background. The utilization of such a technique showed an improved U-Net model performance in medical image segmentation tasks [47,48]. In this work, we used attention blocks in the same manner as originally described in [47]. Blocks were implemented in the decoder part before the skipped connection and upsampled-data concatenation. Attention blocks usually amplify relevant information from the previous decoding layer in image reconstruction (decoder part, skipped connection as in Figure 3) and reduce weights on background features. Implementation is shown in Figure 7. As currently drafted, the output of the attention gate is concatenated with upsampled data from the previous layer.

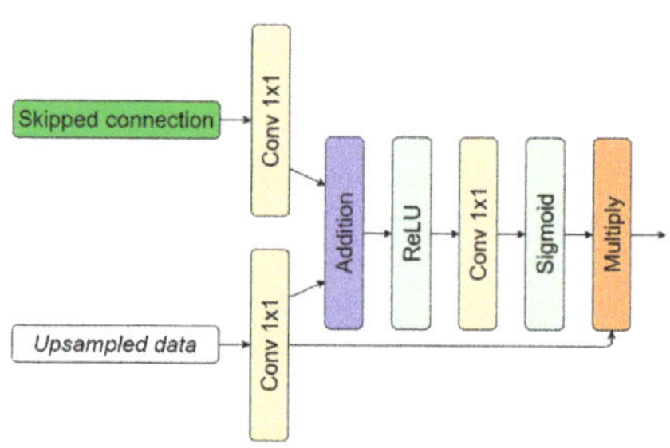

Figure 7. Attention block representation. Upsampled data input taken from the decoder part shown in Figure 2, decoding layer representation before concatenation.

3. Data

3.1. *CrackForest*

The CrackForest [2,3] dataset consists of 118 labeled color images taken with an iPhone 5 camera containing noise: Oil stains, road marking, shoe contours, and shadows. Images are 480 × 320 8 bit RGB. Every image has its ground-truth image with its pixel labeling.

Label mark 1 corresponds to a good surface; 2, crack; 3, surface enclosed by cracks or area is surrounded by cracks; and 4, narrow, hard-to-see cracks. All 118 images had marks 1 and 2, only 22 images contained pixels with labels 3, and label 4 appeared only in 5 images. The image named 042.jpg mismatched with its corresponding mask image. Label 3 marks were debatable in this dataset, given that they were not equally marked in images, and some elements are differently marked from image to image. Therefore, only the two first classes were used in this research. All 117 images were randomly divided into training and testing sets at 70–30%, respectively—82 images for training, and 35 for testing and converted to greyscale. An example of the data sample can be seen in Figure 8.

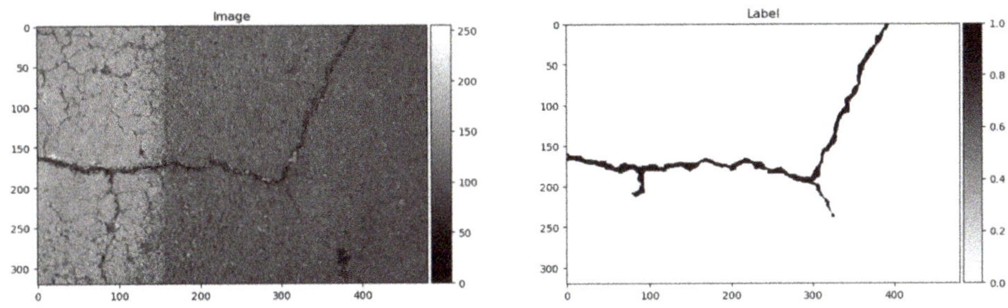

Figure 8. CrackForest data sample. (**a**) Image; (**b**) ground truth.

Multiple researchers from all over the world used this dataset to practice crack detection, and, according to the researchgate.net portal, the citations of the CrackForest database [16] exceed 108 publications. The variety of applied methods goes from simple rule-based methods [49] to moderate image processing by edge detection [50] or superpixel [51] techniques and histogram features [52] to advanced deep learning-based methods for crack detection [53]. Method evaluation differs from author to author by the used evaluation metrics and strategies, depending on article goals. Some authors proposed to use the tolerance distance from 2 to 5 pixels to overcome data-labeling inaccuracy [14]. The best published results are summarized in Table 1.

Table 1. Best results on CrackForest dataset.

Authors	Tolerance in Pixels	Precision	Recall	Dice
Wu et al. [54]	0	0.4330	0.7623	0.4809
Liu et al. [53]	2	0.9748	0.9639	0.9693
Lau et al. [55]	2	0.9702	0.9432	0.9555
Fan et al. [56]	2	0.9119	0.9481	0.9244
Escalona et al. [57]	5	0.9731	0.9428	0.9575

3.2. *Crack500*

The Crack500 dataset was introduced in [18], and it contains images taken using cell phones around the main campus of Temple University. It consists of pixelwise annotated pictures around 2000 × 1500 pixels (varying sizes). It has 250 training, 200 testing, and 50 validation samples. As per the authors in [18], it is the biggest pixelwise annotated road pavement defect dataset. Data samples can be seen in Figure 9.

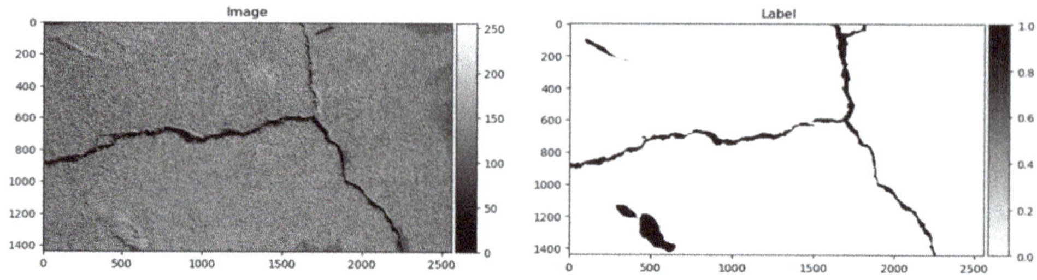

(**a**) (**b**)

Figure 9. Crack500 data sample. (**a**) Image; (**b**) ground truth.

3.3. GAPs384

GAPs384 [18] is a derivation of the German Asphalt Pavement Distress (GAPs) dataset proposed in [20,21]. The original dataset is annotated with bounding boxes, while the modified variant (GAPs384) is labeled pixelwise. GAPs384 is part of the GAPs dataset. It provides HD images (1920 × 1080) with a per-pixel resolution of 1.2 × 1.2 mm. GAPs consists of 353 training and 27 testing samples. Pictures were captured in summer 2015 under dry and warm conditions with a specialized mobile mapping system, S.T.I.E.R of Lehman + Partner GmbH. The imaging system consists of two photogrammetrically calibrated monochrome cameras (1920 × 1080 resolution each), while both cameras covered a single driving lane. The GAPs384 dataset consists of cracks, potholes, and filled cracks. Quite challenging samples can be found that include sewer lids, sidewalk rock fragments, rubbish, and worn-off road lines. The big challenge in a particular dataset is non-uniform illumination through the picture. An example of the dataset can be seen in Figure 10.

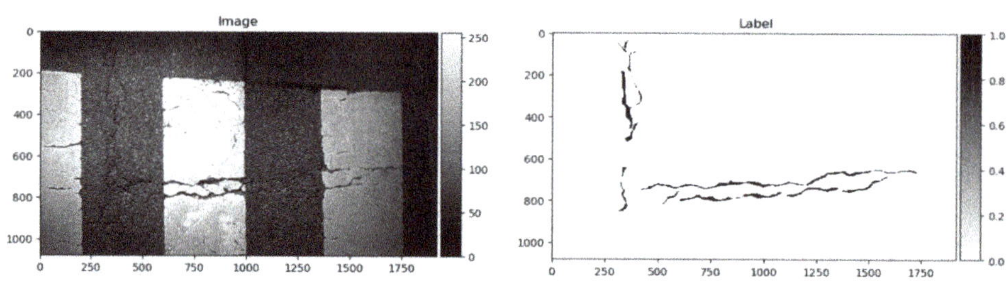

(**a**) (**b**)

Figure 10. German Asphalt Pavement Distress (GAPs) data sample: (**a**) Image; (**b**) ground truth.

3.4. Data Preparation

In this investigation, we took into consideration three different datasets as described above. Every dataset consisted of different-sized pictures. In Crack500, data samples come in different sizes. Moreover, the image size itself is quite big in the Crack500 and GAPs384 datasets, and it is a problem related to the neural network scalability through a limited amount of computer resources (8 GB graphics-card memory in Nvidia 2070 SUPER). Because of these reasons, data samples were cropped into 320 × 320 px patches with 20 px overlap in all datasets. Prepared data were augmented by rotating by 90, 180, and 270°. However, CrackForest in general contained the smallest number of samples compared with the other two. Additional augmentation with flipping and brightness correction in the range of (−15, 15) was introduced to extend the training part of the CrackForest dataset. The size of the Crack500 samples was reduced twice before image cropping to patches due to the extreme size of images compared with the other datasets.

4. Experiments and Evaluation

The neural network algorithm was written in Python (v3.7.4) using Keras API [58] with a Tensorflow 2.0 [59] backend. Experiments were made on a computer with Intel i3 9100F CPU and Nvidia RTX 2070 SUPER 8 GB GPU. Model training and testing were done in a Windows 10 environment.

As described in Section 2, we conducted experiments on several architectures:

- U-Net (Baseline);
- U-Net with residual connections (ResU-Net);
- U-Net with residual connections and ASPP module (ResU-Net + ASPP);
- U-Net with residual connections and ASPP module when connected in "Waterfall" order (ResU-Net + ASPP_WF);
- U-Net with residual connections ASPP and AG modules (ResU-Net + ASPP + AG); and
- U-Net with residual connections ASPP (connected in "Waterfall" order) and AG modules (ResU-Net + ASPP_WF + AG).

In every model training, we picked the combined loss function solution consisting of cross-entropy (Equation (1)) and Dice loss (Equations (2) and (3)). The first part, cross-entropy, is quite often used as a loss function that describes the likelihood of two sets. It can be found in popular machine learning frameworks. Cross-entropy loss is the **X** value relation to the $\dot{\mathbf{X}}$ value in the following expression:

$$L_{CE} = -\frac{\sum_{i=1}^{N} x_i \cdot \log(\dot{x}_i)}{N}, \tag{1}$$

where L_{CE} is the cross-entropy loss; x_i is the ith pixel value in the label matrix **X**; $\dot{x}_i$ is the ith pixel value in the neural network prediction matrix $\dot{\mathbf{X}}$; and N is the total number of pixels.

Another target function is Dice [60] loss. Different than cross-entropy, Dice loss evaluates the overlap of two datasets that are measured in the range from 0 to 1. In image segmentation, the Dice score describes the overlap of sets, label, and prediction:

$$D_{score} = \frac{2 \cdot |\mathbf{X} \cap \dot{X}|}{|\mathbf{X}| + |\dot{X}|}, \tag{2}$$

$$L_D = 1 - D_{score}, \tag{3}$$

where D_{score} denotes the Dice score; **X** is the label matrix; $\dot{\mathbf{X}}$ is the predicted matrix; and L_D is the Dice loss.

The final loss function solution used in this work is expressed in the following equation:

$$L = 0.5L_D + 0.5L_{CE}, \tag{4}$$

where L is the loss function; L_D is the Dice loss; and L_{CE} is the cross-entropy loss.

Datasets used in this investigation might not be fully consistent to make for a generalized pavement crack detector for the majority of the cases. We conducted a few experiments on the smallest dataset, CrackForest. The U-Net model was trained for 50 epochs on the CrackForest training set, and tested on GAP384 data (Figure 11c). The model tended to react sensitively to extraneous objects; in this particular case, sewer lids on the road and non-uniform light (image sides in Figure 11a). A small amount of data as in CrackForest proposes a limited amount of general information that is covered in other datasets (GAPs384, Crack500). As a result, the trained model fails to generalize the global context and is not able to distinguish "unseen" objects from pavement defects, although it might be enough to fit the model to the same dataset (make it perform well in the same dataset's test part). In a few studies [41,55,61], models were pretrained with additional data (or only pretrained the encoding part from segmentation models). The advantage of such weight reuse might be twofold: Faster model training (converging) on the new data, and the ability to improve the generalization and overall prediction performance by introducing more various data with correct labels (as is shown in the comparison table of [45] with models induced with other datasets). We took a similar strategy in this investigation. First, all datasets were mixed for initial network weight training. To equalize every datum from every set, the CrackForest set was additionally augmented by brightness correction and flipping, as was described in Section 3. Models were trained on a mixed dataset for 15 epochs with a 0.001 learning rate at the start, and scheduled reduction by half every 5 epochs. In every epoch, 5636 steps/iteration with a minibatch of 4 were made. Data was shuffled on every epoch start. The model's performance, trained on a mixed dataset, on the testing sample can be seen in Figure 11d. After training with mixed data, every neural network architecture was trained with every dataset individually for 15 additional epochs with a 0.0005 learning rate at the beginning with a reduction by half every 5 epochs. Only values from the neural network output with a higher or equal to 50% confidence rate were taken into consideration. The best performing solution (according to Dice score) from every training were evaluated with accuracy, recall, precision, Dice score (same formula can be expressed as in Equation (2)), and intersection over union (IoU) measures:

$$Accuracy = \frac{TP + TN}{TP + TN + FP + FN}, \quad (5)$$

$$Recall = \frac{TP}{TP + FN}, \quad (6)$$

$$Precision = \frac{TP}{TP + FP}, \quad (7)$$

$$D_{score} = \frac{2 * Precision * Recall}{Precision + Recall}, \quad (8)$$

$$IoU = \frac{GroundTruth \cap Prediction}{GroundTruth \cup Prediction}, \quad (9)$$

where TP is the true positive (correct detection of pixels belonging to labeled defect area); TN is the true negative (nondefective background pixels correctly recognized by detector); FP is the false positive (wrongly detected defect pixels); FN is the false negative (defect pixels undetected by detector); GroundTruth is the labeled image pixels. Precision is the proportion of false alarms; Recall is the proportion of undetected defect pixels; and D_{score} denotes the Dice score or harmonic mean of the precision and recall.

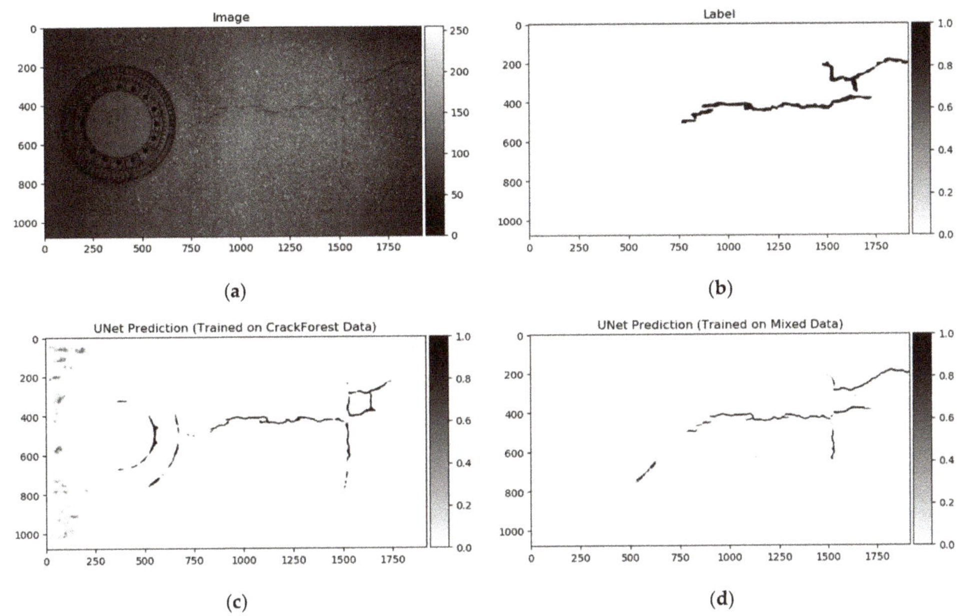

Figure 11. Prediction on the test sample from the GAPs384 dataset. (**a**) Image; (**b**) ground truth; (**c**) full prediction output of the model trained with the CrackForest dataset; (**d**) full prediction output of the model trained with the mixed dataset.

5. Results

As in the previous section, at the beginning, models were trained on the mixed dataset, and individually on every separate dataset. The best-performing solutions of models pretrained and additionally trained on the specific dataset Dice score are shown in Figure 12. Each pretrained model weight was the same for every set.

As shown above, the Dice score was improved (in the case of each dataset) with additional training dedicated to the corresponding dataset. The increase in score might be related to the annotation quality or to the experts' knowledge (that labeled datasets) of problem interpretability. In Figure 8b, Figure 9b, Figure 10b, and Figure 11b show that the details and precision in the sample labels varied. Even in a few annotations from the same dataset (GAPs384, Figures 10b and 11b), the manner of pavement labels might be different. The label in Figure 11b was quite thicker than that in Figure 10b. Training on the mixed dataset in this case possibly ended up fitting the prediction style more or less in favor of one expert (annotation style, such as precision, label line thickness, and other marking properties introduced in specific data sample annotations). Overall, training on the mixed dataset does not highlight a significance in the individual datasets using different architecture neural networks. The increase in Dice was noticeable in most of the cases after short additional training on the individual dataset (Figure 12a–c). Taking U-Net as a baseline for comparison with other models, it is prominent that it has been surpassed by any other architecture. Models induced with a residual connection (ResU-Net) had a slight increase in the Dice score. A more noticeable change can be seen in the GAPs384 dataset that might be related to the data complexity because this dataset introduced more samples with extraneous objects, such as those described in Section 3. Moreover, a big challenge in this particular case is illumination. GAPs384 data require a more powerful model solution for a score increase. Even a larger improvement could be seen by adding an atrous spatial pooling module (ASPP) to the bottleneck of the model. The exact place of addition can be seen in Figure 3. As was described in Section 2, we used two types of links in the ASPP module (Figure 6a,b). Models induced with residual connections and atrous spatial pyramid pooling showed a prediction–performance improvement in the Dice score in all datasets (ResU-Net + ASPP and ResU-Net + ASPP_WF architectures). Models with different types of connections (parallel and Waterfall) in the ASPP part differently favored individual datasets. The biggest increase (from the baseline) could be seen in the GAPs384 set, while the Dice score changed from 0.5448 (U-Net) to 0.5786 (ResU-Net + ASPP). Additionally, we added attention gates (AG) to the ResU-Net + ASPP and ResU-Net + ASPP_WF architectures. However, this enhancement did not always deliver better results. In the CrackForest dataset, the ResU-Net + ASPP + AG and ResU-Net + ASPP_WF + AG neural networks yielded lower results than those of the models without AG modules. On the contrary, the Dice score of the ResU-Net + ASPP + AG architectures surpassed that of the ResU-Net + ASPP in the Crack500 and GAPs384 datasets, and with GAPs384 data, ResU-Net + ASPP + AG achieved the top score.

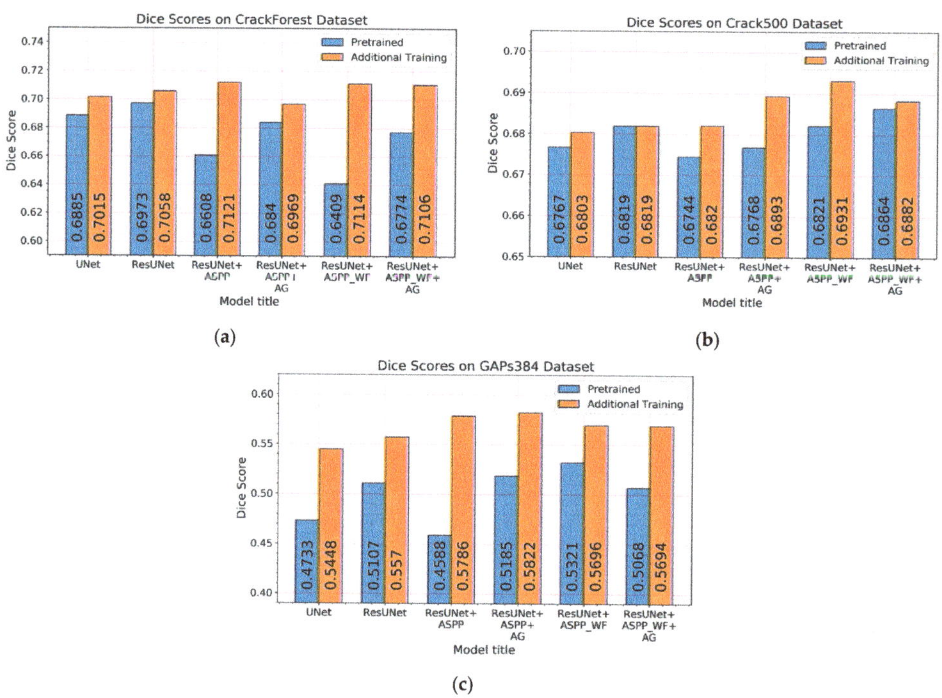

Figure 12. Dice score of every model trained with the mixed dataset (blue bars) and after being trained with a specific dataset (orange bars): (**a**) CrackForest, (**b**) Crack500, (**c**) GAPs384

All measured parameters are given in Table 2, with the highest scores in bold. Accuracy does not represent the prediction performance well because the pavement defects in the image were small, and models performed well by correctly recognizing the background (the biggest part of the image). Intersection over union (IoU) corresponded directly to the Dice score: Models with the highest Dice score delivered the highest IoU. The importance of the false positive (FP) and false negative (FN) costs is described by the recall and precision, respectively. None of these parameters had the top value with U-Net, but in few cases (recall in CrackForest and Crack500 datasets, precision in all datasets), the baseline produced better results than those of some other architecture. Nonetheless, the most important parameter in this investigation was the Dice score, which takes into consideration both recall and precision (as described in Equation (8)).

Table 2. Each model's best weight-prediction results on individual datasets.

CrackForest	Accuracy	Recall	Precision	IoU	Dice
U-Net (Baseline)	0.9898	0.7465	0.6803	0.5489	0.7015
ResU-Net	0.9901	0.7391	0.6928	0.5546	0.7058
ResU-Net+ASPP	0.9902	0.7474	0.692	0.5603	0.7121
ResU-Net + ASPP + AG	0.9899	0.7271	0.6906	0.5442	0.6969
ResU-Net + ASPP_WF	0.9900	0.7494	0.6896	0.5595	0.7114
ResU-Net + ASPP_WF + AG	0.9896	0.7695	0.6715	0.5575	0.7106
Crack500	Accuracy	Recall	Precision	IoU	Dice
U-Net (Baseline)	0.9845	0.7033	0.6996	0.5282	0.6803
ResU-Net	0.9846	0.7002	0.7083	0.5306	0.6819
ResU-Net + ASPP	0.9848	0.6944	0.7152	0.5311	0.6820
ResU-Net + ASPP + AG	0.9841	0.7386	0.6808	0.5389	0.6893
ResU-Net + ASPP_WF	0.9843	0.7524	0.6789	0.5430	0.6931
ResU-Net + ASPP_WF + AG	0.9832	0.7829	0.6447	0.5373	0.6882
GAPs384	Accuracy	Recall	Precision	IoU	Dice
U-Net (Baseline)	0.9953	0.4798	0.7231	0.3925	0.5448
ResU-Net	0.9954	0.4957	0.7134	0.4038	0.557
ResU-Net + ASPP	0.9948	0.5754	0.6285	0.4224	0.5786
ResU-Net + ASPP + AG	0.9951	0.5526	0.6675	0.4264	0.5822
ResU-Net + ASPP_WF	0.9955	0.5459	0.7232	0.4179	0.5696
ResU-Net + ASPP_WF + AG	0.9955	0.5251	0.7143	0.4162	0.5693

Differences in segmentation performance may vary depending on the neural network architectural designs and datasets. The complexity of the analyzed datasets is quite severely altered. Improvements of the Dice scores in different cases are more significant: In GAPs384 between U-Net and ResU-Net + ASPP + AG, and in CrackForest between U-Net and ResU-Net + ASPP. While it can be the main indicator of performance, it is hard to only interpret the segmentation quality from statistical parameters. Representation of the properties, such as the ability to extract a particular feature, for example, narrow defects, can be explained through a visual investigation of the prediction results. As Figures 13–15 show, distinctness in pavement defect extraction is noticeable between the baseline (U-Net) and the best-performing solution. The highlight of the better-performing models is the ability to extract hard-to-see indistinctive cracks that the baseline solution fails to do. As it can be seen in Figures 13c, 14c and 15c, U-Net model falls behind in extremely narrow cracks detection compared with models with residual connection and ASSP module (and AG module in Figure 15d) shown in Figure 13d, Figure 14d, Figure 15d. Segmentation continuity is a feature of better detail extraction. In the bottom of Figure 13c and in whole Figure 14c can be noticed that U-Net architecture cannot make continues pavement crack prediction in more complicated cases, while the best-performing solutions shown in Figures 13d and 14d do segmentation with less flaws. More detailed defect extraction is performed by ResU-Net + ASPP + AG model (Figure 15d) compared with the baseline architeture solution (Figure 15c).

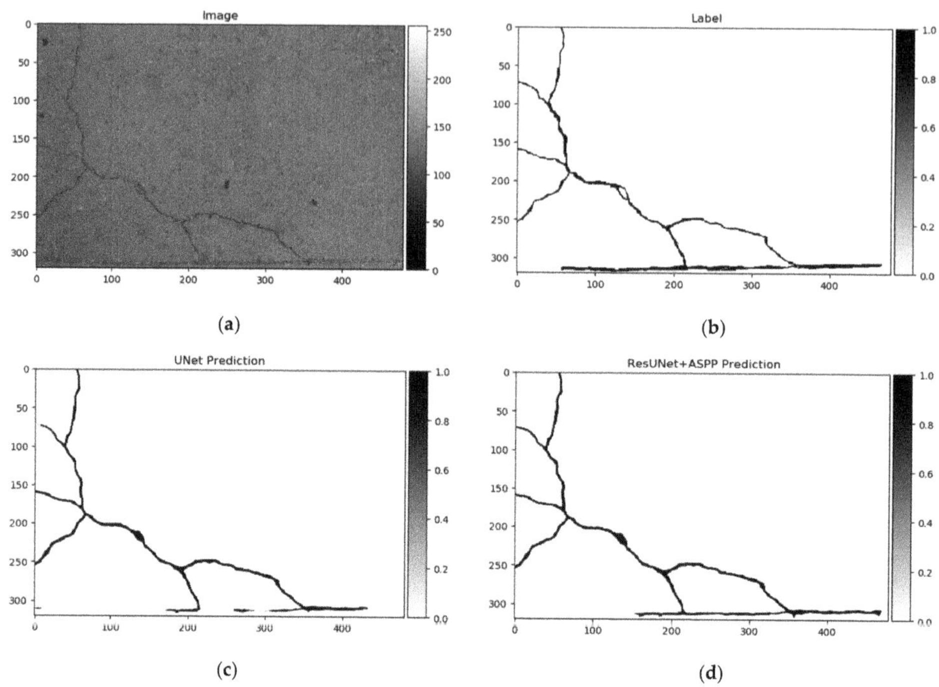

Figure 13. U-Net (baseline) and best-performing solution—ResU-Net + ASPP predictions on CrackForest dataset sample. (**a**) Image, (**b**) label, (**c**) U-Net prediction, (**d**) ResU-Net + ASPP prediction. Segmentation differences significant around image bottom and top left. ResU-Net + ASPP architecture delivered more consistent segmentation.

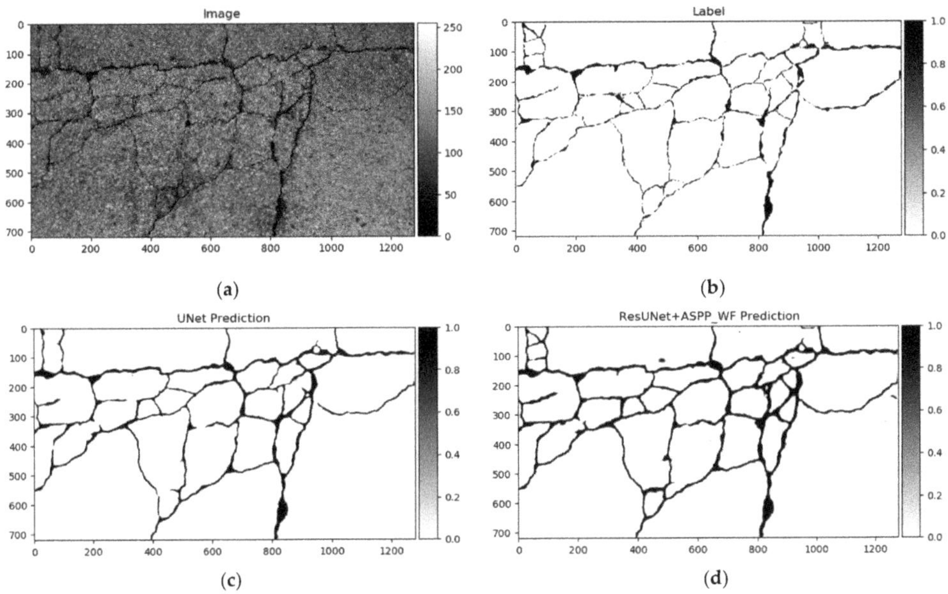

Figure 14. U-Net (baseline) and best-performing solution—ResU-Net + ASPP_WF predictions on Crack500 dataset sample. (**a**) Image, (**b**) label, (**c**) U-Net prediction, (**d**) ResU-Net + ASPP_WF prediction. Quality of continuous defect segmentation noticeable in whole image in predictions (**c**, **d**). ResU-Net + ASPP_WF segmented narrow pavement cracks.

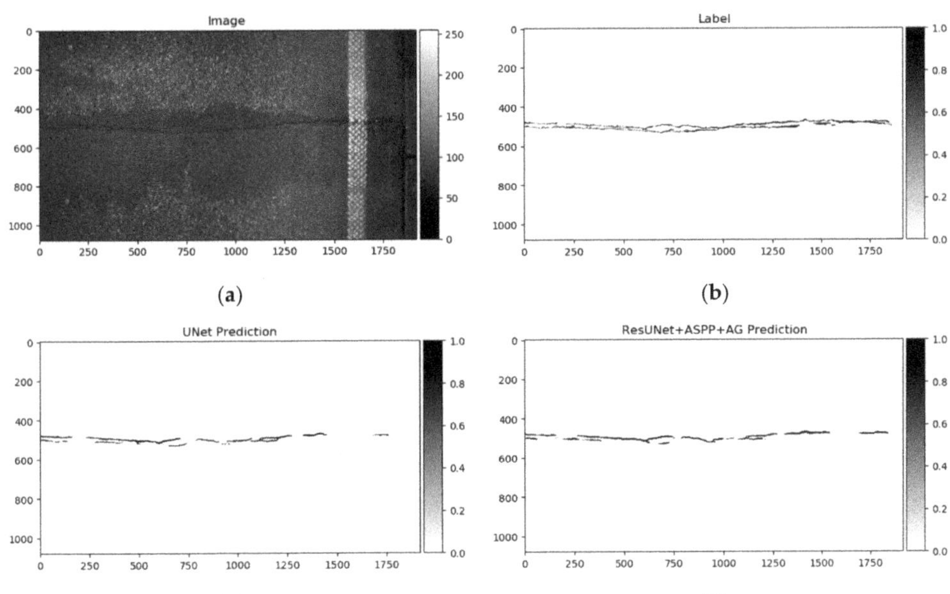

Figure 15. U-Net (baseline) and best performing solution—ResU-Net + ASPP + AG predictions on GAPs384 dataset sample. (**a**) Image, (**b**) label, (**c**) U-Net prediction, (**d**) ResU-Net + ASPP + AG prediction. ResU-Net + ASPP + AG architecture could capture more details compared to U-Net.

While inducing neural networks with additional modules, such as residual connection, ASPP, or AG, we increased the computational complexity (Figure 16a). Extensions enlarged architectures by raising the number of their parameters and by affecting the required time to make the prediction (Figure 16b), taking a bit longer to train the model (Figure 16c). Additional residual connections in U-Net did not make a significant difference, although an increase in the number of parameters was made more than twice by introducing the ASPP module. By adding it in the latent space (bottleneck, Figure 3), we also increased the number of parameters: 256 of 3 × 3 feature kernels in three parallel convolutional operations (Figure 6) for an eightfold downscaled input dimension. However, the number of parameters is not proportional to the computational performance (Figure 16b), and bigger solutions (induced with residual connection and ASPP) took only from 2.55 to 3.27 milliseconds longer to predict in the ResU-Net + ASPP_WF and ResU-Net + ASPP + AG configurations, respectively, compared with U-Net on a 320 × 320 px grayscale-image patch. Inducing models with attention gates did not affect the number of parameters significantly either. Figure 7 shows that it consisted of lightweight 1 × 1 convolutions that did not produce a large computational overhead for the model.

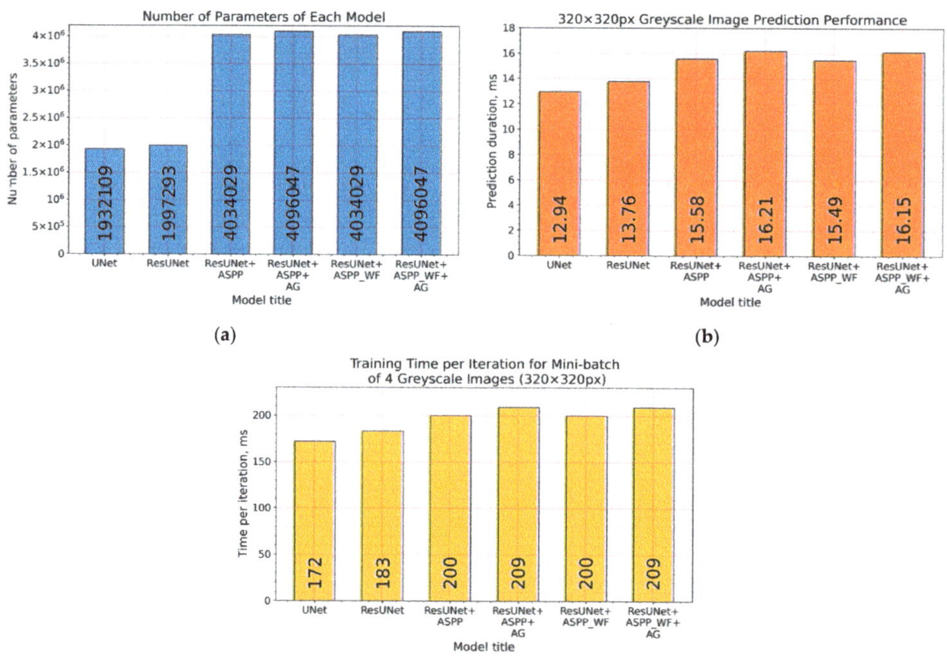

Figure 16. (**a**) Number of parameters, (**b**) prediction computational performance (duration in milliseconds) on a grayscale 320 × 320 px size image, and (c) training time per iteration (duration in milliseconds) for the minibatch of four grayscale 320 × 320 px images of each architecture. Performance evaluation and training made with Nvidia 2070S GPU.

Furthermore, the authors in [14] noted that a certain pixel tolerance can be introduced to cope with annotation inaccuracy. In pavement defect labeling, it is hard to define crack boundaries in a complex pattern. As can be seen in Figure 11a,b, Figure 13a,b, Figure 14a,b, and Figure 15a,b, variations in problem interpretation may seem different in different datasets, and crack label thickness can be subjective (Figure 17). This can cause severe deterioration in statistical performance evaluation, especially considering that a crack itself can be narrow, and its area, compared to the background, is small. We introduced a two- and five-pixel tolerance to the statistical evaluation of the best-performing architecture in each dataset; the results are given in Table 3.

A small allowed error of two pixels significantly boosted the segmentation performance, helping to ignore the slight imprecision appearing on the edge of the label (Figure 17). Increasing the tolerance to up to 5 pixels did not make as big an improvement as that of two pixels, although it might depend on the image resolution and detail complexity. In the dataset, when containing higher-resolution images (Crack500 or GAPs384), the Dice score rise is higher. Comparing our results on CrackForest with scores proposed by another (Table 1), our used solution was in first place with the zero- and five-pixel tolerance, and second with the two-pixel tolerance. It is hard to accurately compare results since the CrackForest dataset is not divided into training and testing parts, and forming them randomly can lead to a specific sample correlation favoring the proposed solution.

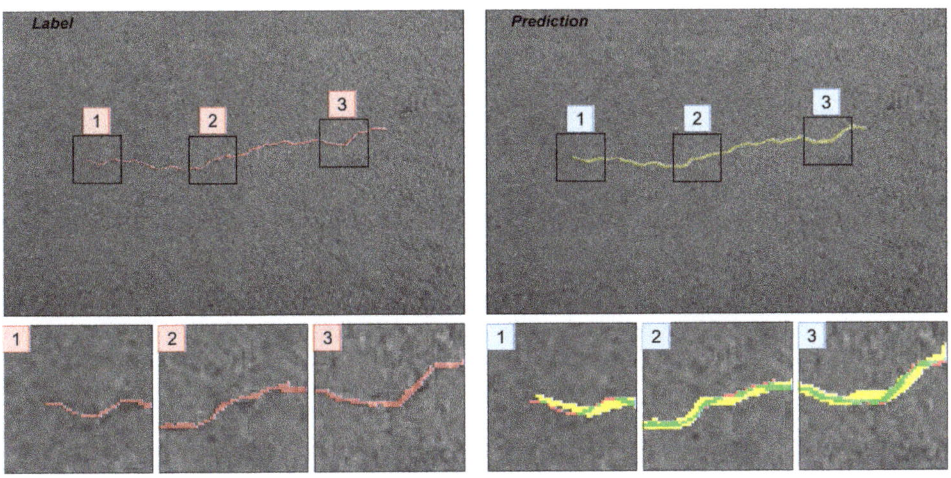

(a) (b)

Figure 17. (**a**) Label and (**b**) prediction of U-Net rendered on images from CrackForest and zoomed regions. Green, overlap of label and prediction; red, prediction pixels; yellow, prediction pixels.

Table 3. Performance evaluation with the 0, 2-, and 5-pixel tolerance on the baseline (U-Net) and best-performing architectural solutions in every dataset.

CrackForest	Tolerance, px	Accuracy	Recall	Precision	IoU	Dice
U-Net (Baseline)	0	0.9898	0.7465	0.6803	0.5489	0.7015
U-Net (Baseline)	2	0.9983	0.9797	0.9194		0.9486
U-Net (Baseline)	5	0.9990	0.9994	0.9411		0.9694
ResU-Net + ASPP	0	0.9900	0.7494	0.6896	0.5595	0.7114
ResU-Net + ASPP	2	0.9986	0.9879	0.9280	-	0.9570
ResU-Net + ASPP	5	0.9991	1.0000	0.9472	-	0.9729
Crack500	Tolerance, px	Accuracy	Recall	Precision	IoU	Dice
U-Net (Baseline)	0	0.9845	0.7033	0.6996	0.5282	0.6803
U-Net (Baseline)	2	0.9957	0.9403	0.8759	-	0.9070
U-Net (Baseline)	5	0.9982	0.9949	0.9323	-	0.9626
ResU-Net + ASPP_WF	0	0.9841	0.7386	0.6808	0.5389	0.6893
ResU-Net + ASPP_WF	2	0.9960	0.9309	0.9017	-	0.9161
ResU-Net + ASPP_WF	5	0.9986	0.9932	0.9481	-	0.9702

Table 3. *Cont.*

GAPs384	Tolerance, px	Accuracy	Recall	Precision	IoU	Dice
U-Net (Baseline)	0	0.9953	0.4798	0.7231	0.3925	0.5448
U-Net (Baseline)	2	0.9979	0.9742	0.6799	-	0.8009
U-Net (Baseline)	5	0.9986	1.0000	0.7772	-	0.8746
ResU-Net + ASPP + AG	0	0.9951	0.5526	0.6675	0.4264	0.5822
ResU-Net + ASPP + AG	2	0.9981	0.9438	0.7280	-	0.8219
ResU-Net + ASPP + AG	5	0.9988	0.9997	0.8127	-	0.8966

6. Discussion

In this paper, we extended and improved our previous work [26] on pixelwise pavement crack detection by using a convolutional neural network. An investigation of road crack segmentation was scaled up by introducing additional datasets, Crack500 and GAPs384. We also demonstrated architectural improvements to the baseline model, U-Net, which boosted the prediction performance. Network structure enhancements with residual connections, atrous spatial pyramid pooling (ASPP), and attention gates (AG) were experimentally trialed on three different and one mixed datasets. In every dataset case model, the configuration with residual connections and ASPP module outperformed U-Net and ResU-Net. Moreover, the Waterfall connection type in the ASPP module did not favor every dataset. The top result with a particular Waterfall ASPP decision was received in the Crack500 data. The model with the AG module only delivered the highest Dice score in the GAPs384 dataset. This architecture (ResU-Net + ASPP + AG) showed the biggest improvement compared to the baseline (U-Net), with a Dice score improvement from 0.5448 to 0.5822, and with a prediction time on a 320 × 320 greyscale image of 12.94 and 16.21 milliseconds, respectively, using Nvidia 2070S GPU. The introduced pixel tolerance significantly boosted the statistics, up to 0.8219 with two pixels and 0.8966 Dice score, with five pixels of allowed error. Visual segmentation inspection revealed that models induced with residual connections and ASPP modules (and AG modules in few cases) tended to capture more complicated details in pavement patterns, and make segmented cracks more continuous.

Neural network training on a mixed dataset and testing on separate datasets did not deliver consistent results with different architectures. Short additional training on the targeted dataset using pretrained (on mixed data) weights gave a better Dice score. Considering that all three datasets were annotated by different experts, the model could tend to fit one or another problem interpretation presented in the labels that might not favor all datasets. From the described data, the annotation style and details varied. Training only on a limited number of samples (as was described by using the model trained with CrackForest on GAPs384) might not be good at generalization.

In a future work, we are considering revising annotations and introducing even more different data for the problem. As collecting and labeling data samples is time demanding and requires precision, synthetic data might also be introduced in the model learning process. While traditional image-processing methods, such as rotation, brightness correction, and noise addition, can be limited in complicated cases, techniques, such as generative adversarial networks (GANs) or variational autoencoders (VAEs), can be engaged to deal with the particular problem. This showed promising results in recent studies [62,63], and it might be a possible solution for the analyzed task.

Author Contributions: Conceptualization, R.A. and A.L.; Data curation, R.A. and A.L.; Investigation, R.A. and A.L.; Methodology, R.A. and A.L.; Resources, R.A. and A.L.; Software, R.A. and A.L.; Supervision, A.L.; Validation, R.A. and A.L.; Visualization, R.A. and A.L.; Writing original draft, R.A. and A.L.; Review and editing, R.A. and A.L. All authors have read and agreed to the published version of the manuscript.

Funding: This research received no external funding.

Conflicts of Interest: The authors declare no conflict of interest.

References

1. Gopalakrishnan, K. Deep learning in data-driven pavement image analysis and automated distress detection: A review. *Data* **2018**, *3*, 28. [CrossRef]
2. ImageNet Large Scale Visual Recognition Challenge (ILSVRC). Available online: http://image-net.org/challenges/LSVRC/ (accessed on 22 December 2019).
3. Hwang, D.; Kim, D.E. Special features on intelligent imaging and analysis. *Appl. Sci.* **2019**, *9*, 4804. [CrossRef]
4. Cao, W.; Liu, Q.; He, Z. Review of pavement defect detection methods. *IEEE Access* **2020**, *8*, 14531–14544. [CrossRef]
5. Sy, N.T.; Avila, M.; Begot, S.; Bardet, J.C. Detection of defects in road surface by a vision system. In Proceedings of the MELECON 2008—The 14th IEEE Mediterranean Electrotechnical Conference, Ajaccio, France, 5–7 May 2008; Volume 2, pp. 847–851.
6. Koch, C.; Brilakis, I. Pothole detection in asphalt pavement images. *Adv. Eng. Inform.* **2011**, *25*, 507–515. [CrossRef]
7. Salari, E.; Bao, G. Automated pavement distress inspection based on 2D and 3D information. In Proceedings of the 2011 IEEE International Conference on Electro/Information Technology, Mankato, MN, USA, 15–17 May 2011 2011; pp. 2–5.
8. Nisanth, A.; Mathew, A. Automated Visual Inspection of Pavement Crack Detection and Characterization. *Int. J. Technol. Eng. Syst.* **2014**, *6*, 14–20.
9. Otsu, N. A Threshold Selection Method from Gray-Level Histograms. *IEEE Trans. Syst. Man Cybern.* **1979**, *9*, 62–66. [CrossRef]
10. Danielsson, P.-E.; Seger, O. Generalized and Separable Sobel Operators. In *Machine Vision for Three-Dimensional Scenes*; Academic Press: San Diego, CA, USA, 1990; pp. 347–3870.
11. Cord, A.; Chambon, S. Automatic Road Defect Detection by Textural Pattern Recognition Based on AdaBoost. *Comput. Civ. Infrastruct. Eng.* **2012**, *27*, 244–259. [CrossRef]
12. Schapire, R.E. A Brief Introduction to Boosting. In Proceedings of the 16th International Joint Conference on Artificial Intelligence—Volume 2; Morgan Kaufmann Publishers Inc.: San Francisco, CA, USA, 1999; pp. 1401–1406.
13. Zhang, L.; Yang, F.; Zhang, Y.; Zhu, Y.J. Road crack detection using deep convolutional neural network. In Proceedings of the 2016 IEEE International Conference on Image Processing (ICIP), Phoenix, AZ, USA, 25–28 September 2016; pp. 3708–3712.
14. Jenkins, M.D.; Carr, T.A.; Iglesias, M.I.; Buggy, T.; Morison, G. A Deep Convolutional Neural Network for Semantic Pixel-Wise Segmentation of Road and Pavement Surface Cracks. In Proceedings of the 2018 26th European Signal Processing Conference (EUSIPCO), Rome, Italy, 3–7 September 2018; pp. 2134–2138.
15. Ronneberger, O.; Fischer, P.; Brox, T. U-net: Convolutional networks for biomedical image segmentation. In Proceedings of the International Conference on Medical Image Computing and Computer-Assisted Intervention, Munich, Germany, 5–9 October 2015; pp. 234–241.
16. Shi, Y.; Cui, L.; Qi, Z.; Meng, F.; Chen, Z. Automatic road crack detection using random structured forests. *IEEE Trans. Intell. Transp. Syst.* **2016**, *17*, 3434–3445. [CrossRef]
17. Cui, L.; Qi, Z.; Chen, Z.; Meng, F.; Shi, Y. *Pavement Distress Detection Using Random Decision Forests*; Springer: Sydney, Australia, 2015; pp. 95–102.
18. Yang, F.; Zhang, L.; Yu, S.; Prokhorov, D.; Mei, X.; Ling, H. Feature Pyramid and Hierarchical Boosting Network for Pavement Crack Detection. *IEEE Trans. Intell. Transp. Syst.* **2020**, *21*, 1525–1535. [CrossRef]
19. Fan, R.; Bocus, M.J.; Zhu, Y.; Jiao, J.; Wang, L.; Ma, F.; Cheng, S.; Liu, M. Road crack detection using deep convolutional neural network and adaptive thresholding. In Proceedings of the 2019 IEEE Intelligent Vehicles Symposium (IV), Paris, France, 9–12 June 2019; pp. 474–479.
20. Eisenbach, M.; Stricker, R.; Seichter, D.; Amende, K.; Debes, K.; Sesselmann, M.; Ebersbach, D.; Stoeckert, U.; Gross, H. How to get pavement distress detection ready for deep learning? A systematic approach. In Proceedings of the 2017 International Joint Conference on Neural Networks IJCNN, Anchorage, AK, USA, 14–19 May 2017; pp. 2039–2047.
21. Stricker, R.; Eisenbach, M.; Sesselmann, M.; Debes, K.; Gross, H. Improving Visual Road Condition Assessment by Extensive Experiments on the Extended GAPs Dataset. In Proceedings of the 2019 International Joint Conference on Neural Networks (IJCNN), Budapest, Hungary, 14–19 July 2019; pp. 1–8.

22. Wu, S.; Zhong, S.; Liu, Y. Deep residual learning for image steganalysis. *Multimed. Tools Appl.* **2018**, *77*, 10437–10453. [CrossRef]
23. Li, B.; Wang, K.C.P.; Zhang, A.; Fei, Y.; Sollazzo, G. Automatic Segmentation and Enhancement of Pavement Cracks Based on 3D Pavement Images. *J. Adv. Transp.* **2019**, *2019*, 1813763. [CrossRef]
24. Loupos, K.; Doulamis, A.D.; Stentoumis, C.; Protopapadakis, E.; Makantasis, K.; Doulamis, N.D.; Amditis, A.; Chrobocinski, P.; Victores, J.; Montero, R.; et al. Autonomous robotic system for tunnel structural inspection and assessment. *Int. J. Intell. Robot. Appl.* **2018**, *2*, 43–66. [CrossRef]
25. Protopapadakis, E.; Voulodimos, A.; Doulamis, A.; Doulamis, N.; Stathaki, T. Automatic crack detection for tunnel inspection using deep learning and heuristic image post-processing. *Appl. Intell.* **2019**, *49*, 2793–2806. [CrossRef]
26. Augustaukas, R.; Lipnickas, A. Pixel-wise Road Pavement Defects Detection Using U-Net Deep Neural Network. In *Proceedings of the 2019 10th IEEE International Conference on Intelligent Data Acquisition and Advanced Computing Systems: Technology and Applications (IDAACS), Metz, France, 18–21 September 2019*; IEEE: Metz, France, 2019; pp. 468–472.
27. Augustaukas, R.; Lipnickas, A. Road Pavement Segmentation Project Code. Available online: https://github.com/rytisss/RoadPavementSegmentation (accessed on 22 April 2020).
28. He, K.; Zhang, X.; Ren, S.; Sun, J. Delving deep into rectifiers: Surpassing human-level performance on imagenet classification. In Proceedings of the IEEE International Conference on Computer Vision (ICCV), Santiago, Chile, 7–13 December 2015; pp. 1026–1034.
29. Ioffe, S.; Szegedy, C. Batch Normalization: Accelerating Deep Network Training by Reducing Internal Covariate Shift. *arXiv* **2015**, arXiv:1502.03167.
30. Van Laarhoven, T. L2 Regularization versus Batch and Weight Normalization. *arXiv* **2017**, arXiv:1706.05350.
31. Chu, Z.; Tian, T.; Feng, R.; Wang, L. Sea-Land Segmentation With Res-UNet And Fully Connected CRF. In Proceedings of the IGARSS 2019—2019 IEEE International Geoscience and Remote Sensing Symposium, Yokohama, Japan, 28 July–2 August 2019; pp. 3840–3843.
32. Xiao, X.; Lian, S.; Luo, Z.; Li, S. Weighted Res-U-Net for High-Quality Retina Vessel Segmentation. In Proceedings of the 2018 9th International Conference on Information Technology in Medicine and Education (ITME), Hangzhou, China, 19–21 October 2018; pp. 327–331.
33. Riid, A.; Lõuk, R.; Pihlak, R.; Tepljakov, A.; Vassiljeva, K. Pavement distress detection with deep learning using the orthoframes acquired by a mobile mapping system. *Appl. Sci.* **2019**, *9*, 4829. [CrossRef]
34. Xu, W.; Liu, H.; Wang, X.; Qian, Y. Liver segmentation in CT based on ResU-Net with 3D Probabilistic and Geometric Post Process. In Proceedings of the 2019 IEEE 4th International Conference on Signal and Image Processing (ICSIP), Wuxi, China, 19–21 July 2019; pp. 685–689.
35. He, K.; Zhang, X.; Ren, S.; Sun, J. Deep Residual Learning for Image Recognition. In Proceedings of the 2016 IEEE Conference on Computer Vision and Pattern Recognition (CVPR), Las Vegas, Nevada, 26 June–1 July 2016; pp. 770–778.
36. Yu, F.; Koltun, V. Multi-Scale Context Aggregation by Dilated Convolutions. *arXiv* **2015**, arXiv:1511.07122.
37. Chen, L.C.; Papandreou, G.; Kokkinos, I.; Murphy, K.; Yuille, A.L. DeepLab: Semantic Image Segmentation with Deep Convolutional Nets, Atrous Convolution, and Fully Connected CRFs. *IEEE Trans. Pattern Anal. Mach. Intell.* **2018**, *40*, 834–848. [CrossRef]
38. Chen, L. Rethinking Atrous Convolution for Semantic Image Segmentation. *arXiv* **2017**, arXiv:1706.05587.
39. Liu, W.; Rabinovich, A.; Berg, A.C. ParseNet: Looking Wider to See Better. *arXiv* **2015**, arXiv:1506.04579.
40. Wang, Y.; Liang, B.; Ding, M.; Li, J. Dense semantic labeling with atrous spatial pyramid pooling and decoder for high-resolution remote sensing imagery. *Remote Sens.* **2019**, *11*, 20. [CrossRef]
41. Chen, G.; Li, C.; Wei, W.; Jing, W.; Woźniak, M.; Blažauskas, T.; Damaševičius, R. Fully convolutional neural network with augmented atrous spatial pyramid pool and fully connected fusion path for high resolution remote sensing image segmentation. *Appl. Sci.* **2019**, *9*, 1816. [CrossRef]
42. Zhang, P.; Ke, Y.; Zhang, Z.; Wang, M.; Li, P.; Zhang, S. Urban land use and land cover classification using novel deep learning models based on high spatial resolution satellite imagery. *Sensors* **2018**, *18*, 3717. [CrossRef] [PubMed]
43. Bo Guo, Y.; Matuszewski, B.J. Giana polyp segmentation with fully convolutional dilation neural networks. In Proceedings of the 14th International Joint Conference on Computer Vision, Imaging and Computer Graphics Theory and Applications, Prague, Czech Republic, 25–27 February 2019; Volume 4, pp. 632–641.

44. Chen, L.-C.; Zhu, Y.; Papandreou, G.; Schroff, F.; Adam, H. Encoder-Decoder with Atrous Separable Convolution for Semantic Image Segmentation. *arXiv* **2018**, arXiv:1802.02611.
45. Artacho, B.; Savakis, A. Waterfall atrous spatial pooling architecture for efficient semantic segmentation. *Sensors* **2019**, *19*, 5361. [CrossRef]
46. Jetley, S.; Lord, N.A.; Lee, N.; Torr, P.H.S. Learn To Pay Attention. In Proceedings of the ICLR 2018, Vancouver, Canada, 30 April–3 May 2018.
47. Schlemper, J.; Oktay, O.; Schaap, M.; Heinrich, M.; Kainz, B.; Glocker, B.; Rueckert, D. Attention gated networks: Learning to leverage salient regions in medical images. *Med. Image Anal.* **2019**, *53*, 197–207. [CrossRef]
48. Oktay, O.; Schlemper, J.; Folgoc, L.L.; Lee, M.; Heinrich, M.; Misawa, K.; Mori, K.; McDonagh, S.; Hammerla, N.Y.; Kainz, B.; et al. Attention U-Net: Learning Where to Look for the Pancreas. *arXiv* **2018**, arXiv:1804.03999.
49. Cheng, H.D.; Chen, J.-R.; Glazier, C.; Hu, Y.G. Novel Approach to Pavement Cracking Detection Based on Fuzzy Set Theory. *J. Comput. Civ. Eng.* **1999**, *13*, 270–280. [CrossRef]
50. Saar, T.; Talvik, O. Automatic Asphalt pavement crack detection and classification using Neural Networks. In Proceedings of the 2010 12th Biennial Baltic Electronics Conference, Tallinn, Estonia, 4–6 October 2010; pp. 345–348.
51. Stutz, D.; Hermans, A.; Leibe, B. Superpixels: An evaluation of the state-of-the-art. *Comput. Vis. Image Underst.* **2018**, *166*, 1–27. [CrossRef]
52. Velinsky, S.A.; Kirschke, K.R. Design Considerations for Automated Pavement Crack Sealing Machinery. In Proceedings of the Second International Conference on Applications of Advanced Technologies in Transportation Engineering, Minneapolis, Minnesota, 18-21 August 1991; pp. 77–80.
53. Liu, W.; Huang, Y.; Li, Y.; Chen, Q. FPCNet: Fast Pavement Crack Detection Network Based on Encoder-Decoder Architecture. *arXiv* **2019**, arXiv:1907.022481.
54. Wu, S.; Fang, J.; Zheng, X.; Li, X. Sample and Structure-Guided Network for Road Crack Detection. *IEEE Access* **2019**, *7*, 130032–130043. [CrossRef]
55. Lau, S.L.H.; Wang, X.; Xu, Y.; Chong, E.K.P. Automated Pavement Crack Segmentation Using Fully Convolutional U-Net with a Pretrained ResNet-34 Encoder. *arXiv* **2020**, arXiv:2001.01912.
56. Fan, Z.; Wu, Y.; Lu, J.; Li, W. Automatic Pavement Crack Detection Based on Structured Prediction with the Convolutional Neural Network. *arXiv* **2018**, arXiv:1802.02208.
57. Escalona, U.; Arce, F.; Zamora, E.; Sossa, H. Fully convolutional networks for automatic pavement crack segmentation. *Comput. Sist.* **2019**, *23*, 451–460. [CrossRef]
58. Keras. Available online: https://keras.io/ (accessed on 22 December 2019).
59. Tensorflow. Available online: https://www.tensorflow.org/ (accessed on 22 December 2019).
60. Dice, L.R. Measures of the Amount of Ecologic Association Between Species. *Ecology* **1945**, *26*, 297–302. [CrossRef]
61. Iglovikov, V. TernausNet: U-Net with VGG11 Encoder Pre-Trained on ImageNet for Image Segmentation. *arXiv* **2018**, arXiv:1801.05746.
62. Popescu, D.; Ichim, L.; Stoican, F. Flooded Area Segmentation from UAV Images Based on Generative Adversarial Networks. In Proceedings of the 2018 15th International Conference on Control, Automation, Robotics and Vision (ICARCV), Singapore, 18–21 November 2018; pp. 1361–1366.
63. Liu, F.; Xia, Y.; Yang, D.; Yuille, A.; Xu, D. An Alarm System For Segmentation Algorithm Based On Shape Model. In Proceedings of the IEEE International Conference on Computer Vision (ICCV), Seoul, Korea, 27 October–2 November 2019.

sensors

MDPI

Article

A Cost-Effective Electric Vehicle Intelligent Charge Scheduling Method for Commercial Smart Parking Lots Using a Simplified Convex Relaxation Technique

Muhammad Jawad [1], Muhammad Bilal Qureshi [2,*], Sahibzada Muhammad Ali [2], Noman Shabbir [3], Muhammad Usman Shahid Khan [4], Afnan Aloraini [5] and Raheel Nawaz [6]

1 Department of Electrical and Computer Engineering, CUI Lahore Campus, Lahore 54000, Pakistan; mjawad@cuilahore.edu.pk

2 Department of Electrical and Computer Engineering, CUI Abbottabad Campus, Abbottabad 22060, Pakistan; hallianali@cuiatd.edu.pk

3 Department of Electrical Power Engineering & Mechatronics, Tallinn University of Technology, 19086 Tallinn, Estonia; noshab@taltech.ee

4 Department of Computer Science, CUI Abbottabad Campus, Abbottabad 22060, Pakistan; ushahid@cuiatd.edu.pk

5 Department of Computer Science, Qassim University, Al Qassim 1162, Saudi Arabia; A.ALOURANI@qu.edu.sa

6 Department of Operations Technology, Events and Technology Management, Manchester Metropolitan University, Manchester M15 6BH, UK; R.Nawaz@mmu.ac.uk

* Correspondence: bilalqureshi@cuiatd.edu.pk

Received: 13 June 2020; Accepted: 26 July 2020; Published: 27 August 2020

check for updates

Abstract: Deployment of efficient and cost-effective parking lots is a known bottleneck for the electric vehicles (EVs) sector. A comprehensive solution incorporating the requirements of all key stakeholders is required. Taking up the challenge, we propose a real-time EV smart parking lot model to attain the following objectives: (a) maximize the smart parking lot revenue by accommodating maximum number of EVs and (b) minimize the cost of power consumption by participating in a demand response (DR) program offered by the utility since it is a tool to answer and handle the electric power usage requirements for charging the EV in the smart parking lot. With a view to achieving these objectives, a linear programming-based binary/cyclic (0/1) optimization technique is developed for the EV charge scheduling process. It is difficult to solve the problems of binary optimization in real-time given that the complexity of the problem increases with the increase in number of EV. We deploy a simplified convex relaxation technique integrated with the linear programming solution to overcome this problem. The algorithm achieves: minimum power consumption cost of the EV smart parking lot; efficient utilization of available power; maximization of the number of the EV to be charged; and minimum impact on the EV battery lifecycle. DR participation provide benefits by offering time-based and incentive-based hourly intelligent charging schedules for the EV. A thorough comparison is drawn with existing variable charging rate-based techniques in order to demonstrate the comparative validity of our proposed technique. The simulation results show that even under no DR event, the proposed scheme results in 2.9% decrease in overall power consumption cost for a 500 EV scenario when compared to variable charging rate method. Moreover, in similar conditions, such as no DR event and for 500 EV arrived per day, there is a 2.8% increase in number of EV charged per day, 3.2% improvement in the average state-of-charge (SoC) of the EV, 12.47% reduction in the average time intervals required to achieve final SoC.

Keywords: intelligent charging; demand response; electric vehicle; linear programming; optimization; smart parking; smart grid

1. Introduction

The conventional transportation system is producing nearly 14% of total worldwide greenhouse discharges, which is estimated to increase further 50% by 2030 [1]. Air pollution alone is one of the major extraneous costs of transportation, especially as it directly influences the health of local inhabitants. The growing international desire for adopting environment-friendly technologies has resulted in the acceptance and usage of alternate fuel vehicles that can be hybrid and battery-operated electric vehicles (EVs). However, EVs have been unable to secure the confidence of customers and acquire a large market share yet. This is mainly due to their technical and infrastructure limitations, such as limited driving range and unavailability of charging facilities. Moreover, the resultant cost considerations also deter potential customers. In order to further the public acceptance of EVs, it is important to improve the EV infrastructure for all key stakeholders.

A significant aspect of these infrastructure requirements is the abundant availability of charging points and charging stations. Charging places act as connection sources to the electrical grid for the EV, and the place of powering up the batteries for EV drivers. However, a major challenge for existing power systems will be to maintain demand response (DR) with the growing demand for electricity as a result of an increased number of EV parking lots. Currently, how to manage the adaptability of EV use and their charging without having a drastic impact on existing power grids is a contentious topic [2,3]. The recent advancements in control strategies and sensing visualize DR as an effective tool to help address the issue of demand-supply mismatch in electric power grids. The DR program provides customers with leverage to shift their electricity demand from peak hours to off-peak hours and as a consequence, derive benefits in terms of lower electricity prices and financial incentives. In Refs. [4–6], a number of DR programs were explored to lower the overall power usage of a household for load shifting and curtailing the appliances of residential households during peak hours.

Modern utility grids need suitable candidates that can be involved in the DR programs for demand curtailment of the grid in peak hours. The large-scale EV parking lots are one of the most suitable solution due to following reasons: (a) by managing the accumulated electrical load of the parking lot can act as a large-scale demand curtailment player in the power market and (b) the electrical load flexibility of EV charging loads with adjustable and interruptible features can ensure that EV charge scheduling can coordinate with the grid's DR programs to satisfy all charging demand of EVs, while fulfilling committed demand curbing for DR. In DR programs, the electricity tariffs are usually dynamic in nature and may include day-ahead market pricing [7,8], real-time pricing [9,10] or time-of-use structure of pricing [11]. Due to the adjustable features present in the EV charging load, it is possible for EV smart parking lots to explore possibilities to execute both incentive-based and price-based DR programs.

Therefore, in this paper, a real-time charge management system (CMS) is developed to charge EVs parked in a smart parking lot by taking into consideration the advantages of DR programs. The EV charging mechanism is optimized using the linear programming (LP) and convex relaxation techniques. The objective of the CMS is to maximize the number of EVs that can be fully charged over a 24-h period and minimize the cost of energy consumption by participating in different DR programs. Moreover, the DR event will also curtail the total charging load of the smart parking lot. In the proposed CMS, the on-off (binary) charging strategy is employed to solve the EV charging problem while considering the state-of-health (SoH) of the EV battery and its C-rate. The C-rate is a measure of the rate at which a battery is charge/discharged relative to its maximum capacity. Under the binary charging technique, the EV attached to the charging pole will be charged at fixed maximum charging rate as per C-rate equals to 1 C. Thus, in addition to consuming less time, it will also have minimal impact on the battery health by fully utilizing the battery's single charging cycle. However, the optimization of binary charging decision for every EV is a non-trivial task because the scheduling of EVs is formulated as a binary optimization task. The intelligent charging schedule can be determined using the generate and test method, also known as exhaustive search. However, it is not recommended for real-time implementation due to its computational expensive nature. Therefore, the problem is mitigated by developing a convex relaxation approach that is integrated with the LP, where near intelligent charging scheduling is computed. Finding the intelligent charging schedule for EV is a non-convex problem due to the high number of EV, and random arrival and departure times. Therefore, the binary constraint in the problem related to the on-off charging strategy is relaxed to solve the problem as a convex problem using LP. Thereafter, a modified mapping is used to convert the solution back to binary values. This whole process is named as the simplified convex relaxation approach integrated with the LP. More details about this process have been provided in Section 5. The main contributions of the paper are as follows:

Modelling of a realistic EV smart parking lot by considering: (a) regular and random arrival and departure times for EV; (b) different types of EV along with different battery capacities and charging rates; (c) level-2 battery charging standard; and (d) EV battery efficiency and life cycle.

- First, an on-off (0/1) charging technique is used for the EV, i.e., an EV selected for charging in a time interval will be charged at fixed maximum charging rate according to its level II charge rate. Charging an EV at constant power may extend the battery's service time. Secondly, the communication overheads will be small as only a small subset of the EV will be required to contact and it would be more feasible to apply on/off charging scheme. However, the target is to charge maximum EV with the minimum charging price possible. Whereas, the scheduling of the EV for intelligent charging using the on-off/cyclic technique is a binary optimization issue that is computationally extensive to be resolved at run-time if the number of EV will increase. In order to address this issue, an EV charge scheduling technique named simplified convex relaxation approach integrated with the LP is utilized.
- The DR program can lower the cost of electricity for EV parking lots; therefore, DR events are introduced in the on-off charging technique. The on-off charging scheme not only respond to the variable electricity prices but also responds well for demand curtailment events from the grid.
- Second, a variable charging rate technique for the EV charging is tested while having fixed capacity limit of the EV charging station, i.e., all the charging poles in the EV parking lot are used to accommodate all arrived EVs. However, the drawbacks of variable charging rates are: (a) charging EVs at constant power could extend the service time of the battery, which is a disadvantage in variable charging rate [12] (b) variable charge rate will extend the charging time of the battery.
- A thorough comparison of both binary charging scheme and variable charging scheme is conducted based on maximum revenue generation, energy consumption cost, number of EVs charging in a daytime, and the impact on the battery life of the EVs.

The paper is organized as follows: A detailed literature review is presented in Section 2, the system model and optimization technique are discussed in Section 3, Section 4 comprises of results, discussions, and comparative analysis. Finally, the paper is concluded in Section 5 along with some proposals for future research directions.

2. Literature Review

In the past, researchers have developed EV smart parking lot models and applied numerous optimization techniques in order to schedule EV charging. The objectives of all such EV scheduling techniques in EV smart parking lots are at least one of the following: (a) maximize the number of EVs charged in an allocated time [13]; (b) maximize the smart parking lot profit [14]; (c) minimize the EV owner's charging cost [15]; and (d) minimize the peak demand by participating in DR [16].

In Refs. [13], the authors address the problem of EV charging schedules from the perspective of smart parking lot operator and EV owners. Through the utilization of quadratic problems, the optimization objectives were to maximize the number of EVs charged and the overall revenue of the smart parking lot while minimizing the EV charging costs for the EV owners. Similarly, in Ref. [17], game theory is used to schedule the EV charging to maximize the utilization of the smart parking lot by increasing the number of EVs to be charged. Accordingly, a greater number of EV owners can be accommodated. However, the authors have not considered the stochastic features of electricity price variations and the driving patterns of the EV. In Refs. [18,19], MILP and fuzzy linear programming (FLM) are used, respectively, to maximize the smart parking lot profit by effectively optimizing the charging schedule of the EVs. Similarly, in Ref. [20], a linear programming (LP) technique and dynamic programming (DP) model are used, respectively, to maximize the smart parking lot profit and to minimize the charging costs of EV owners. The problem with LP is how to simultaneously handle both real number and integers; therefore, MILP is more suitable then LP. The DP lacks a general formulation and every problem need to be addressed in its own way. Moreover, the DP consumes more memory while storing the results of intermediate steps, which is not the case in MILP.

The DR program is an effective tool to minimize the cost of energy consumption while participating at a different level of programs. Therefore, in Ref. [14], the authors introduced the participation of EV smart parking lots in incentive-based and price-based DR programs along with the stochastic programming optimization technique, with a view to maximizing the EV smart parking lot profit. A similar technique with same the objective is used in Ref. [21] wherein the authors introduce smart parking lot as an aggregator agent in the real-time DR market. The energy source is parked EVs in the smart parking lot. Therefore, in Ref. [22], the EV smart parking lot is introduced as a multi-energy system (MES) to enhance the profit of the smart parking lot and to improve the operational capability of the MES.

Meanwhile, a graded control algorithm for EV charging across several aggregators is proposed in [11] to minimize the peak load of the smart parking lot and electricity cost. The heuristics are developed in such a manner that at first, the distribution system operator (DSO) will solve the charging curve by participating in the DR program, after which the charging power will be allocated to each EV. An algorithm based on quadratic programming (QP) is proposed in Ref. [15] to optimize the charging schedule of EVs in order to minimize the charging cost of EV owners. Similarly, the DP is used in Ref. [23] for EV charging schedules with an aim to maximize the smart parking lot profit and minimize the EV owner's charging cost. In Ref. [23], the DP is solved by assuming the parked EVs in the smart parking lot as an aggregated battery bank. A simulation platform named Okeanos is proposed in Ref. [24] based on a multi-agent DR program with an aim to get benefits from the optimal EV charging. According to the authors, increasing the number of charged EVs and minimizing the electricity tariff through the DR program can help optimize the EV charging schedule [24]. In Ref. [25] the authors developed an algorithm to minimize the EV owner's charging cost by combining a distributed DR method and parked EVs as a storage capability. Similarly, using the LP algorithm, a distributed DR method using the random usage pattern of EVs is proposed in Ref. [26]. The objective of the model is to minimize the peak demand of the utility grid to minimize electricity cost. In Ref. [27], a real-time EV charging scheme for EV smart parking lot is proposed using MILP that coordinates and priorities requirements of EV charging and discharging powers with the power generation of the utility grid, renewable energy sources (RES), energy storage system (ESS), and electricity price preferences.

In Ref. [28], a distributed EV charge management scheme is proposed from the perspective of EV owners to minimize the wait time of EV charging on parking lots. The authors proposed a P/S communication framework to utilize charging reservation effectively. In Ref. [29], another similar work, the authors proposed a preempted charging recommendation system for the income EV using V2V based reservation system with an aim to minimize on-the-move charge time and travelling time. In our work, we incorporate the arrival and departure time of the EV to ensure no delay in service for EV charging, while the advance reservation system for EVs does not lie in the scope of our work.

To the best of our knowledge and analyzing the previous studies described in this Section, the proposed work is comprehensive in that the proposed scheme aims to use EV smart parking lots as a service provider and a decision-maker in DR program to optimize the intelligent charging schedule of parked EV. We developed an objective function that maximizes the number of EVs charged at a given time. However, the selection process of EV charging involves EV charging priority, state-of-charge (SoC), and electricity pricing preference; therefore, the advantages of the objective function are manifold, such as maximization of the smart parking lot profit, minimization of the EV charging cost for the EV owners, minimization of the peak demand by participating in different DR programs, and minimized impact on the state-of-health (SOH) of the EVs' battery by charging the battery at the maximum and fixed maximum charge rate as per level-II charging of EV at C-rate equals 1 C.

3. System Model

The system model for charging EVs in the EV smart parking lot is depicted in Figure 1. The model includes the following: (a) main grid (utility grid); (b) the aggregated electric load of parked EV; (c) charge management model (CMS); and (d) real-time DR power market. The primary source of energy for the EV smart parking lot is the main grid. The LP and simplified convex relaxation techniques are used to optimize the EV' charge scheduling.

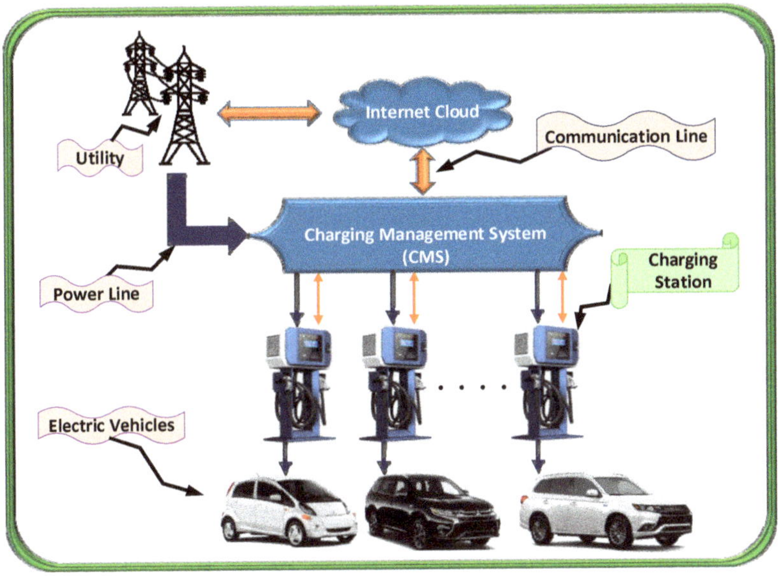

Figure 1. Charging Management System (CMS).

3.1. Preliminary Discussion

In the future, as the demand for EV charging increases, smart parking lot operators will be encouraged to install more charging points to incorporate more EVs. However, charging power capacity is a hard constraint that will limit the number of EVs to be charged at a given point in time. Moreover, variables, such as arrival time, departure time, EV battery capacity, driving cycle, and maximum charge rate of any EV depend on the vehicle type and manufacturer. Therefore, an optimized EV charging strategy is needed to control and manage the charging capacity of the smart parking lot in order to provide charging services to a large number of customers by taking into consideration the departure time of the EVs. Moreover, for the purpose of avoiding maximum capacity overload, a charging priority needs to be determined for the parked EVs to ensure fair charging preferences and to provide enough charging for an EV before departure time. Charging all the parked EVs simultaneously can overburden the utility grid due to the maximum power consumption capacity cap imposed on the smart parking lot. However, by using an intelligent EV charge scheduling algorithm, the peak load demand can be managed, delayed, or optimized by taking into consideration the departure time of the EV. This technique can make EV smart parking lots good candidates as aggregated agents in the DR program. Therefore, the EV smart parking lot will be suitable for the day-ahead electricity pricing tariff by managing their electric load throughout 24-h. Considering all the aforementioned attributes, the proposed CMS is designed in a manner that can optimize the EV charging schedule by managing the request of the DR program using a day-ahead electricity pricing tariff.

As illustrated in Figure 1, the CMS establishes two-way uninterruptable communication between the utility grid and the smart parking lot by sharing the electricity load curve with the grid and optimizing the scheduling of the parked EVs in accordance with the DR requirement of the grid. Moreover, the CMS computes the total electricity consumption cost. Therefore, the objective of the CMS is to manage the DR curve for the smart parking lot. The CMS stores the arrival time of each EV parked at the charging station to optimize the charging schedule of each parked EV. Moreover, the driver needs to enter the departure time and charge ranking on the smart parking station pole. Furthermore, the CMS computes the charge rate for each parked EV as the maximum charging rate varies for different make and model of the EV. Therefore, the idea behind the optimized CMS is to charge the parked EVs at fixed charge rates or halt the charging considering the departure time of the EVs. In this process, there must be an increase in the count of the number of fully charged EV during the whole day.

The motivation to participate in the DR program initiated by the utility grid is the demand flexibility of the charging loads (EVs) in the smart parking lot. The DR programs can be either fixed or time varying. The demand reduction curve programmed by the smart parking lot relies on the DR proceedings that include the fruitful bids processed in the day-ahead environment for the bidding of the required demand profile.

3.2. EV Charge Scheduling Technique

Let P denote the number of charging poles to offer charging facility for the EVs that have arrived in the EV smart parking lot. The CMS will develop a real-time (0/1) cyclic optimized charging schedule technique for the EVs attached to the P charging poles. The arrival and departure of the EVs in the smart parking lot is a continuous process; however, a sampling interval S_t is taken for the decision-related propose of the real-time algorithm. The CMS will optimize the EV charging schedule after every S_t time interval. Therefore, the 24 h in a day are divided into X time intervals such that $X = T/S_t$. If an EV is attached to the pth charging pole at a real-valued time interval t_p^{arr}, an enrolling status will appear, and the charging pole is considered as being triggered. A binary variable δ_p^x will determine the connection state for each pth charging pole at the xth time interval, where p represents the number of charging pole $p = 1, \ldots, P$ and x is the number of time interval $x = 1, \ldots, X$. The variable $\delta_p^x = 1$ if the pth charging pole is connected to an EV; otherwise $\delta_p^x = 0$. Given that the daytime is divided into X time intervals, the arrival time interval of the EV attached to the pth pole is computed as: t_p^{arr}/S_t. A rounding with ceiling operator $\lceil . \rceil$ is applied to select the lowest integer value as $\lceil n_p^{arr} = t_p^{arr}/S_t \rceil$. The driver is bound to insert the departure real-valued time. Similarly, let the departure time provided by the driver at the pth charging pole be t_p^{dep}. The departure time interval of the EV attached to the pth pole is computed as: $\lfloor n_p^{dep} = t_n^{dep}/S_t \rfloor$. The floor rounding operator $\lfloor . \rfloor$ is used to select the highest previous integer value. Meanwhile, a binary variable k_p^x is used to record the charging status of the attached EV. If an EV attached to the pth pole in time interval x is charging, the variable $k_p^x = 1$; otherwise $k_p^x = 0$. However, if $\delta_p^x = 1$, then the k_p^x can have values 0 or 1 depending on whether or not EV is charging, but if $\delta_p^x = 0$, then definitely the $k_p^x = 0$. All the notations used in the system model are defined in Table 1.

As stated earlier, the purpose of the EV charging scheme is to maximize the number of EVs that have been charged during the course of the entire day and to minimize the electricity cost paid to the utility grid by participating in the DR program. Moreover, a valid charging priority scheme is integrated with the optimization problem along with an electricity preference price.

Table 1. Notations used in the System Model.

Notation	Definition
Indices	
p	Index for charging pole changing from $1, \ldots, P$
x	Index for the number of time interval changing from $1, \ldots, X$
I	Index for all charging poles connected with EV at time interval x
Π^x	charging poles indices with respect to charging priority in descending order
Constants	
P	Total number of charging poles
S_t	Sampling Interval
X	Total Number of Time Intervals
t_p^{arr}	Arrival time of an EV attached to pole p
n_p^{arr}	Real-valued arrival time interval of the EV attached to pole p
t_p^{dep}	Departure time of an EV attached to pole p
n_p^{dep}	Real-valued departure time interval of the EV attached to pole p
γ_p^x	Current State-of-Charge of the EV attached to the pole p
$\gamma^{min}, \gamma^{max}$	Minimum and maximum boundary limit for State-of-Charge
C_p^{cap}	Maximum charging capacity of the EV's battery
U_p^x	Total time intervals required for the charging of EV attached to pole p
ω_p	Rank function value of every EV attached to the pole p
Z_p^{max}	Maximum charging rate of an EV attached to the pole p
β_{min}, β_{max}	Lowest and highest electricity rates
Z_{total}	Total power capacity bound on EV parking lot
Z_p^{max}	Maximum charging power drawn by the EV attached to the pole p
C_p^{cap}	Maximum energy storing capacity of the EV attached to the pole p
η	Charging efficiency of the EV
Binary Variables	
δ_p^x	1, if an EV is attached to pole p at time interval x and 0 otherwise
k_p^x	1, if an EV is attached to pole p at time interval x is charging and 0 otherwise
Continuous Variables	
r_p^x	Weighted charging priority of EV attached to pole p at time interval x
α_p^x	Preference level of pth charging pole to charge attached EV
β^x	Electricity rate at each time interval x
Z_{DR}^x	Demand Curbing of demand response event
$\tilde{k}_p^x$	Real-valued charging decision variable

3.3. Electric Vehicle Charging Priority and Preference

A weighting parameter is formulated in order to compute the charging priority of an EV attached to the p-th charging pole. The parameter contains the capacity of the p-th pole to refill the EV battery and the remaining time to charge the battery. The EV battery's state-of-charge (SoC) attached to the p-th pole in time interval x is denoted as γ_p^x and C_p^{cap} represents its battery capacity. Therefore, the charging time intervals required to charge the EV is computed as follows at the x-th time interval:

$$U_p^x = n_p^{dep} - x. \tag{1}$$

We define a variable rank function $\omega_p \in [0, 1]$ for every EV that comes for charging at the smart parking lot. Although a higher rank will be given to the executive customers, they will be paying higher membership fees. Therefore, the weighted charging priority of every EV attached to the pth charging pole during the xth time interval is computed as follows, irrespective of its been charged or not:

$$r_p^x = \begin{cases} \frac{\omega_p C_p^{cap}\left(\gamma^{max} - \gamma_p^x\right)}{Z_p^{max} U_p^x}, \ if \ \delta_p^x = 1; \\ 0, \ if \ \delta_p^x = 0. \end{cases} \tag{2}$$

In Equation (2), the term $\omega_p C_p^{cap}\left(\gamma^{max} - \gamma_p^x\right)$ represents the battery capacity that needs to be filled during the stay of the parked EV up to a maximum limit of the SoC γ^{max}. This term also implies that an EV with a lower SoC will have greater needs for charging. In the proposed work, if an EV attached to the p-th charging pole is selected to be charged, then it will be charged at the maximum charging rate Z_p^{max} of that EV. Therefore, the denominator term of the Equation (2) denotes the maximum charging energy that is provided to the EV. If the value of U_p^x is low, the EV must be charged on priority before the possible departure time. The charging priority in Equation (2) is a normalized factor ($r_p^x \in [0,1]$) because the nominator is divided by a maximum value. If an EV departs from the smart parking lot after the charging process, then the p-th charging pole is set to be free by setting $\delta_p^x = 0$ along with weighted charging priority factor $r_p^x = 0$. The charging pole will be re-activated again by setting $\delta_p^x = 1$, if another EV arrives and re-attaches to the p-th charging pole and its corresponding weighted charging priority factor r_p^x will be computed again using Equation (2).

It is necessary to manage the charging demand of the parked EVs within the desired time period; however, maximizing the smart parking lot profit is another important factor that must not be disregarded. Therefore, maximizing the electricity bill by utilizing dynamic electricity pricing is another important factor. Moreover, the maximum number of EVs should be charged at the time of low electricity pricing to minimize the power consumption cost. In order to model these factors, we assume that β_{max} and β_{min} are the highest and lowest rates of electricity, respectively quoted by the utility grid for the EV smart parking lot. Therefore, the α_p^x is defined as an additional parameter to represent the preference level of the pth charging pole to charge the associated EV for the electricity rate β^x at the xth time interval. The parameter is defined as:

$$\alpha_p^x = \frac{(\beta_{max} - \beta^x)}{(\beta_{max} - \beta_{min})}, \ \forall \, p = 1 \ldots P, \ x = 1 \ldots X. \tag{3}$$

In Equation (3), the value of the parameter α_p^x will be high when the electricity price is low and vice versa. The denominator term is used to normalize the preference level as $\alpha_p^x \in [0,1]$.

4. Optimization Technique for Cyclic (On-Off) Scheduling of Electric Vehicles

The purpose of the EV' charge scheduling technique is to manage the charging of all attached EVs while keeping the maximum power consumption of the smart parking lot under maximum permissible demand limit. The EVs selected for charging at any time interval x is based on the weighted charging priority parameter r_p^x and the preferred electricity rate parameter α_p^x. We defined a set $I = \{1, \ldots, P\}$ to record the indices number of all the charging poles connected to an EV at the time interval x while compiling a set $\varnothing^x = \left\{r_1^x, r_2^x, \ldots, r_P^x\right\}$ that contains the record of the weighted charging priority parameters of every EV attached to the charging poles at the x-th time interval. The purpose of this record is to evaluate each charging pole one by one to observe the preference level of each attached EV and prioritizing the EV with a high weighted charging priority number, while keeping the total charging demand below the maximum available capacity limit, including the curtailment of DR demand.

The purpose of the proposed optimization technique is to find the optimized set of EV to be charged at xth time interval. Therefore, a descending order operator ($Sort(.)$) is applied to the set $\varnothing^x$ along with its indices set I. The sorted charging pole indices with regard to the descending weighted charging priority parameters are stored in a new set Π^x. Therefore, we assign the highest priority to the charging pole having the highest r_p^x value because the charging poles indices set Π^x are rearranged in the descending order and the poles will be selected for charging until the limit of power capacity is reached. The set Π^x is defined as:

$$\Pi^x = Sort(I)\big|_{\varnothing^x} \tag{4}$$

The optimization of the EV charge scheduling is to enhance the number of EVs charged in a given time interval. At the present q-th time interval, an objective function needs to be maximized that is described as the product of weighted charging priority parameters r_p^y and electricity price preference level α_p^y. The expression for the objective function is formulated as:

$$\max_{k_p^y, x=q,..,X} \sum\nolimits_{y=x}^{X} \sum\nolimits_{p\in\Pi^x} k_p^y r_p^y \alpha_p^y \tag{5}$$

In Equation (5), the binary variable k_p^x represent the binary parameters to be optimized. The ideal EV charge scheduling is intended to increase the count of the EVs selected for charging at the given time interval as well as to reduce the electricity cost depending on the variable electricity rates governed by the utility. While the EV attached to the p-th charging pole of the smart parking lot will occupy the charging pole as per time intervals $x \in \left[n_p^{arr}, n_p^{dep}\right]$, the optimization of the EV charging arrangement can be performed from the present q-th time interval till the day ends, such that $x \in [q, X]$ to ease the calculation. This is because it was evident that $r_p^x = 0,\ \forall\, x \in \left[n_p^{dep}, X\right]$, as mentioned in Equation (2).

Suppose Z_{total} is the total power capacity bound offered to the smart parking lot for the EV charging process, excluding the DR program, and let Z_{DR}^x be the demand curbing for the DR event at the given x-th time interval. The overall charging requirement is controlled by the demand boundary limit $\left(Z_{total} - Z_{DR}^x\right)$ at the given x-th time interval, such as:

$$\sum_{p\in\Pi^x} k_p^x Z_p^{max} \le Z_{total} - Z_{DR}^x,\ x = q,\ldots,X. \tag{6}$$

It is ensured that the minimum energy requirement of an EV for the next travelling is fulfilled. Therefore, the charging state for the EV being charged by the p-th charging section at the given x-th time interval is controlled by the lower boundary γ^{min}. Moreover, the SoC of the EV is constrained by the upper boundary γ^{max} to control overcharging. Assume η denotes the EV charging efficiency; therefore, the constraints applied on the charging is written as follows:

$$\gamma_p^x C_p^{cap} + \eta Z_p^{max} k_p^x S_t \ge \gamma^{min} C_p^{cap},\ x = q \ldots X; \tag{7}$$

$$\gamma_p^x C_p^{cap} + \eta Z_p^{max} k_p^x S_t \le \gamma^{max} C_p^{cap},\ x = q \ldots X; \tag{8}$$

Finally, the SoC of each EV's battery for the next time interval is calculated as:

$$\gamma_p^{x+1} = \gamma_p^x + \frac{\eta Z_p^{max} k_p^x S_t}{C_p^{cap}} \tag{9}$$

5. Simplified Convex Relaxation Methodology

The binary optimization function is derived in Equation (5) under the inequality constraints Equations (6)–(8) and SoC update rule defined in Equation (9). In Equation (5), the binary variables include $k_p^x \in \{0,1\}$, $p = 1\ldots P$, $x = 1\ldots X$; therefore, solving Equation (5) is a binary optimization problem. However, the nature of binary search optimization is extensive and influenced by the imprecation of dimensionality. Moreover, binary search is more complicated than linear search as it overkills for a very small number of variables/elements or provides an infeasible solution for an oversized set of variables, such as if the number of time slots X or the number of charging pole P. Furthermore, the list of the variables needs to be sorted to use the binary search algorithm, which is often unfeasible, specifically for the case when the number of variables is constantly increasing. In addition, the binary search algorithm only works for less-than inequality constraints. Therefore, based on multiple testing, we concluded that the binary optimization problem defined in Equations (5)–(8) is not solvable in real-time for the EV charging schedules. To address the aforesaid issue a simplified convex relaxation technique is applied in this paper. The simplified convex relaxation technique is a two-step process. First, the decision variable $k_p^x \in \{0,1\}$ of Equation (5), which is binary in nature, is now relaxed $\hat{k}_p^x \in [0,1]$ to allow the decision for the charging is real-valued. Therefore, the LP is used to optimize the approximated newly defined optimization problem as follows:

$$\max_{\hat{k}_p^x, x=q,..,X} \sum_{y=x}^{X} \sum_{p\in\Pi^x} \hat{k}_p^y r_p^y \alpha_p^y \tag{10}$$

The constraints for the objective function are defined as follows:

$$\sum_{p\in\Pi^x} \hat{k}_p^x Z_p^{max} \leq Z_{total} - Z_{DR}^x,\ x = q \ldots X. \tag{11}$$

$$\gamma_p^x C_p^{cap} + \eta Z_p^{max} \hat{k}_p^x S_t \geq \gamma^{min} C_p^{cap},\ x = q \ldots X; \tag{12}$$

$$\gamma_p^x C_p^{cap} + \eta Z_p^{max} \hat{k}_p^x S_t \leq \gamma^{max} C_p^{cap},\ x = q \ldots X; \tag{13}$$

Computationally, LP is an efficient technique and does not diverge with the increase in system dimensionality and size. Therefore, we found LP more suitable for the scheduling of EV charging in real-time in our application as all variables are linear [30]. However, a problem arises as to the relaxed variable $\hat{k}_p^x$ have fractional value output after LP optimization. With regard to the proposed on-off charging strategy, for EV charging needs to convert back the fractional values of the $\hat{k}_p^x$ to binary values 0 and 1. Therefore, in second step. A precise mapping is used to convert $\hat{k}_p^x \in [0,1]$ to $k_p^x \in \{0,1\}$. Moreover, the constraints defined in Equations (6)–(8) also need to be ensured by the mappings. In order to map fractional values to binary values, the charging poles indices set $\Omega = \{1,\ldots,P_y\}$ is computed such that all the indices of pole having associated $\hat{k}_p^x$ values between 0 and 1 are included in set Ω, where $P_y \leq P$. Moreover, a new set is defined $\Gamma^x = \{\hat{k}_1^x, \hat{k}_2^x, \ldots, \hat{k}_P^x\}$ to store the values of the variable k_p^x for all the charging poles at the x-th time interval. The indices set Ω and variable set Γ^x are used to apply descending order operation on the set Γ^x, such as:

$$\Lambda^x = Sort(I)\big|_{\Gamma^x} \tag{14}$$

In order to generate k_p^x all such poles having $\hat{k}_p^x = 1$ are automatically selected in the matrix k_p^x. The remaining assignment of binary values in the matrix k_p^x involves the satisfaction of Equation (15).

$$\sum_{p\in\Lambda^x} \hat{k}_p^x Z_p^{max} \leq\ Z_{total} - Z_{DR}^x,\ x = q,\ldots,X. \tag{15}$$

At each time interval x, the SoC of each EV is calculated using Equation (9) once the k_p^x is computed.

6. Results and Discussions

6.1. Simulation Settings

In our simulations, we used $P = 200$ charging poles in the EV smart parking lot in order to estimate the effectiveness of the proposed work. The overall time period T is said to be 24 h for the EV charging and sampling time to measure the best and ideal EV charging scheduling, wherein S_t is fixed to 10 min. Therefore, the overall time intervals are given by $X = T/S_t = 144$ min. The SoC for every EV must have a defined higher (γ^{max}) and lower (γ^{min}) limits, selected as 0.99 and 0.66, respectively [31]. Moreover, the day-ahead electricity pricing tariffs of Houston (TX, USA) for the year 2019 are used in the simulations [30].

EV arrival at the smart parking lot can be regular or random in nature; therefore, both types of EVs are taken into consideration in the simulations to present more accurate designing of the EV smart parking lot, where the ratio for former to later is 6:4. The EVs having a regular presence in the smart parking lot can be easily predicted owing to their comparatively fixed entering and leaving times. A typical Gaussian model layout is used to generate synthetic data for both the entry and exit time of the regular EV. The arrival time of the regular EV is generated using normal distribution having mean (μ_{arv}) and standard deviation (σ_{arv}) as 6:00 am and 60 min, respectively, and is given by:

$$f(x) = \frac{1}{\sigma\sqrt{2\tau}}e^{-\frac{1}{2}(\frac{x-\mu}{\sigma})^2} \tag{16}$$

The leaving times of the EV are also generated using the normal Gaussian distribution, where μ_{dep}, and σ_{dep} are 5:00 pm and 120 min, accordingly. In the wake of unforeseeable visits, the random EV have greater uncertain habits of utilizing the EV charging station. In order to simulate the irregular consumption habits, the arrival and departure time of random EV are equally distributed across the scheduling interval $x \in [1, X]$ and the probability density function for gaussian uniform distribution is given by:

$$f(x) = \begin{cases} \frac{1}{2\sigma\sqrt{3}} \; for - \sigma\sqrt{3} \le x - \mu \le \sigma\sqrt{3} \\ 0 \; otherwise \end{cases} \tag{17}$$

Suppose E is the number of EVs arriving at the smart parking lot for charging during the whole day which is, for the intervals $x = 1, \ldots, X$. It is possible that any charging pole can be utilized more than once to charge EVs throughout the day. Therefore, the charging pole can be reused instantly after the admitted car finishes charging and departs from the smart parking station. The simulation is conducted to model and observes different frequencies of EV arrival and for this purpose we used the following settings for the number of EVs entering the smart parking lot during the whole day $E = [100\text{–}500]$ with a step of 100. However, the smart parking lot has limited charging poles that are selected as $P = 200$; for this reason, the variable count of EV are simulated for the time interval $x = 1, \ldots, X$. The probability density function of a different number of EV (E) over the whole day is plotted in Figure 2, which illustrates the occupied charging poles in each time interval. Notably, as soon as the E increases to 300 and above, the charging station seems full for many time intervals given that all 200 charging poles are occupied.

The electrical characteristics of recently manufactured EV and their usage in percentage in the simulations are given in Table 2 [32]. Moreover, the level 2 battery charging mechanism is adopted in the simulations [33]. The initial SoC of an arrived EV is considered as being equally distributed in the interval [0.1–0.4], whereas charging efficiency η of each EV is set to be 0.9 [34]. The amount of EV with membership levels of low, average, and high is set to be 20%, 50% and 30% of total numbers of EV (E), respectively. The overall capacity threshold for the entire smart parking lot is set as $Z_{total} = 500$ kW. All simulations of our proposed model are carried out on a server SYS-7047GR-TRF system using MATLAB optimization toolbox.

Table 2. Battery Capacity, Charge Rate and Division of Electric Vehicles.

EV Types	Battery Capacity (kWh)	Max. Charging Rate at Level II (kW)	Total Percentage of Cars (%)
Mitsubishi i-MiEV	16	3.6	10
Chevy Volt	18	3.6	10
Ford Focus Electric	23	6.6	15
Fiat 500E	24	6.6	15
Kia Soul EV	27	6.6	15
Mercedes B-Class	28	10	5
BMW i-3	33	7.7	5
Volkswagen E-Golf	36	7.2	10
Nissan LEAF	40	6.6	10
Tesla Model-S	100	10	5

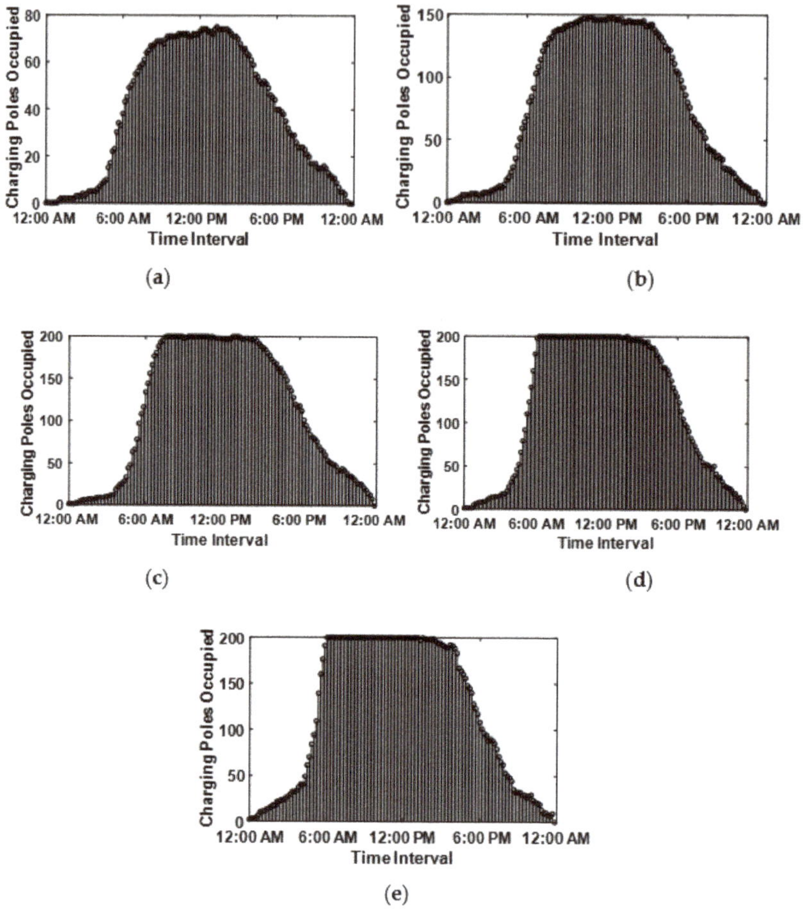

Figure 2. Number of Charging Pole Captured when (**a**) $E = 100$, (**b**) $E = 200$, (**c**) $E = 300$, (**d**) $E = 400$, and (**e**) $E = 500$.

6.2. Simulation Results of Proposed On-Off Charging Scheme

The goal of the proposed algorithm is to fully charge any EV that arrives at the smart parking lot. Since the SoC of each EV is observed before departure; therefore, the SoC of randomly selected five EVs is plotted in Figure 3. In Figure 3, it can be observed that the EVs leaving the smart parking lot are fully charged ($SoC = 1$). In addition, it is evident that the SoC of the EVs is gradually increasing with charging in each time interval x. Similarly, In Figure 4, the consumed and the remaining power of the grid is plotted for the scenario when $E = 500$. The Figure 4 provides evidence that during the working hours, the overall capacity threshold for the entire smart parking lot reaches its limit and many newly arrived EV must wait in a queue before getting an opportunity to be charged. Moreover, the available power is underutilized for many time intervals. Furthermore, the Figure 5 shows that the overall charging cost of the entire smart parking lot is computed separately in each time interval using the number of EV that have been charged in that slot multiplied with the given day-ahead electricity tariff. Moreover, we used flexible charging rates for EVs in this paper given that the day-ahead pricing tariff is also variable. This mode of flexible charging is known as C-F mode [35].

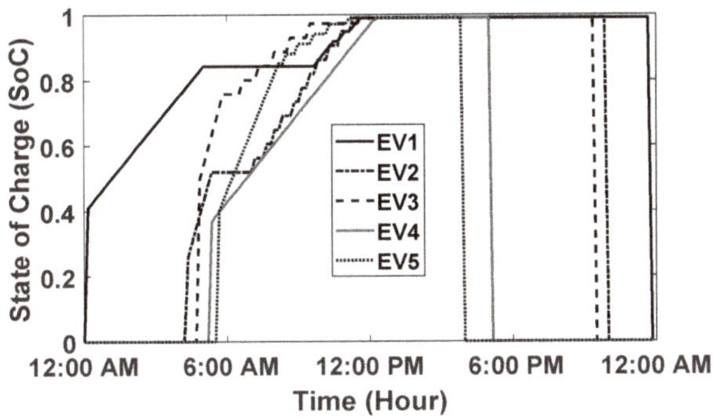

Figure 3. State of Charge of Randomly Selected EV.

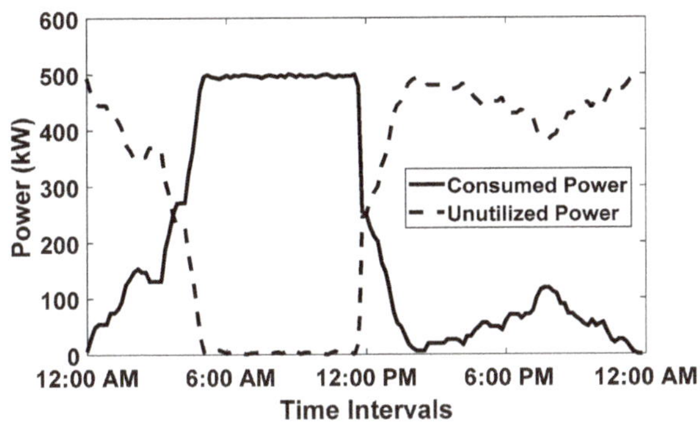

Figure 4. Consumed and Unutilized Power in 24 h, when $E = 500$.

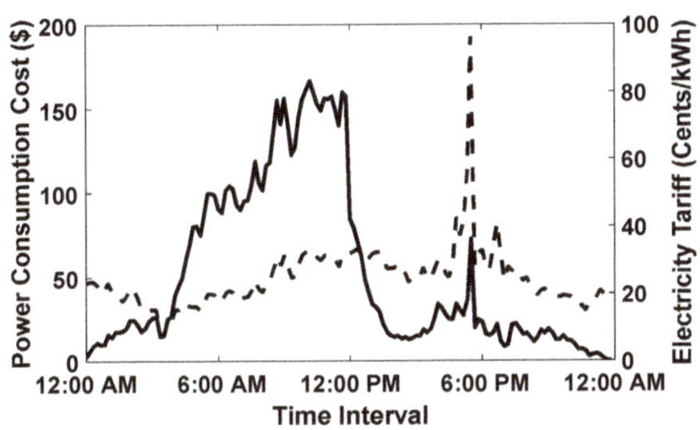

Figure 5. Power Consumption Cost and Electricity Tariff in 24 h, when $E = 500$.

6.3. Simulation Results of Proposed Charging Scheme under Demand Response Events

In order to validate the performance of the proposed On-Off EV charge scheduling technique, three DR events are generated and simulated with the same variable day-ahead electricity tariff illustrated in Figure 5 [21–23]. The first DR event (DR1) is a long load shedding DR event generated from 4:00 pm to 10:00 pm, the simulation results of which are presented in Figure 6a. The second DR event (DR2) comprises of short load shedding events from 6:00 am to 9:00 am and from 5:00 pm to 9:00 pm. The simulation results of the second DR event are illustrated in Figure 6b. Meanwhile, the third DR event (DR3) contains two load shedding events that are variable in time, the simulation results of which are depicted in Figure 6c. The results illustrated in Figure 6a–c shows that for all three

DR events, the EV charge scheduling arrangement does not exceed the maximum power capacity threshold limit for the total number of EVs (*E*) equals 500. Moreover, the inclusion of a parameter α_p^x presented in Equation (3) curtails the optimization process from scheduling the charging of EVs in the high rate tariff intervals to minimize the overall charging cost. Therefore, the charging cost of all three DR events was compared with no DR event and depicted in Figure 7, which shows that under the influence of load shedding events, the overall charging cost for the smart parking lot has not increased in spikes.

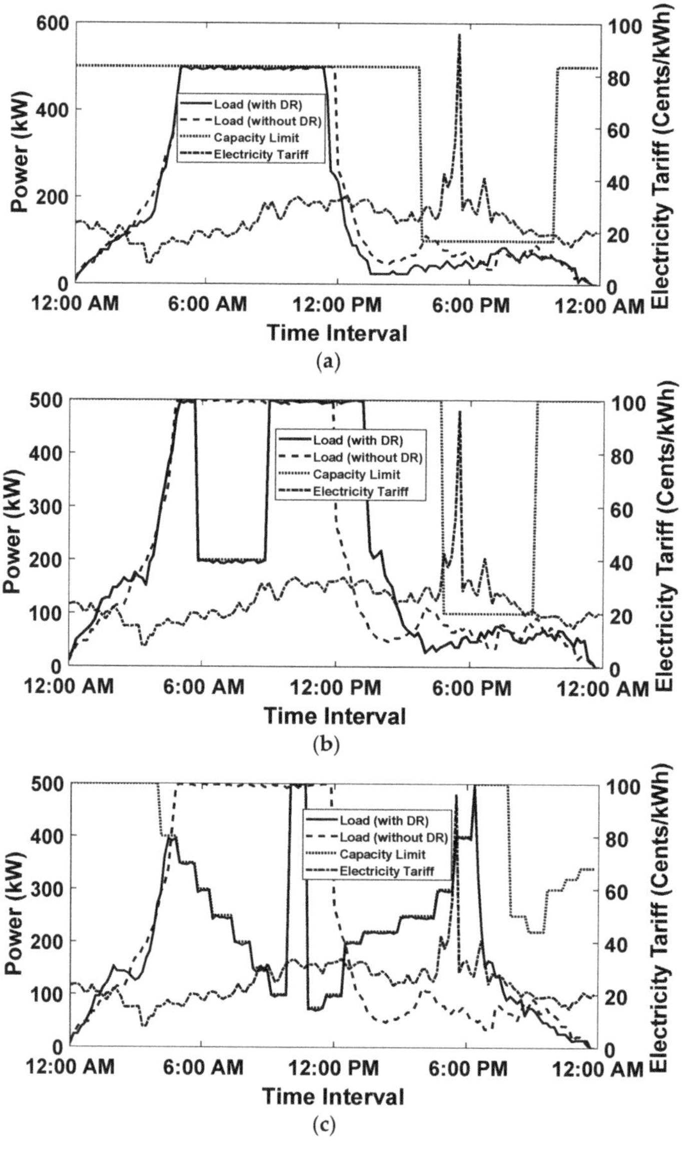

Figure 6. Electric Load Profiles of Smart parking lot with and out DR events. (**a**) Single Constant Load Shedding DR Event (DR1), (**b**) Two Constant Load Shedding DR Event (DR2), (**c**) Variable Load Shedding DR Event (DR3).

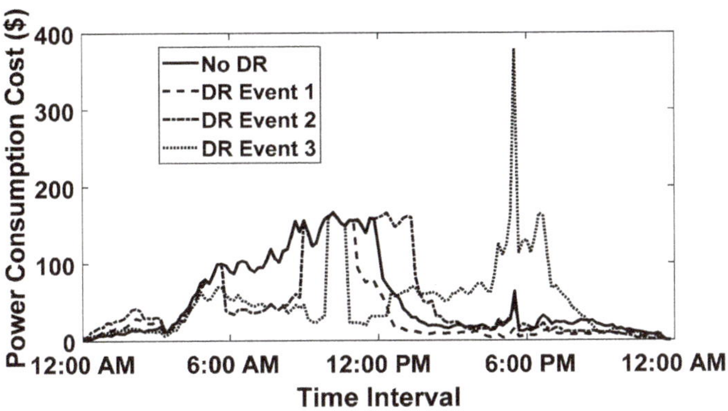

Figure 7. Power Consumption Cost comparison between DR Events.

The proposed charging mechanism is compared against the real-valued variable charging power approach in order to verify the capability and effectiveness of the proposed technique. In variable charging rate technique, the objective is to vary the charge rate of an EV attached to the charging pole instead of charging the EV on fixed charging power, so that the charging efficiency of the EV battery must not be affected in a given time interval and the maximum number of arrived EV can be accommodated in the smart parking lot. This condition would be helpful if an empty smart parking pole was available, but we were waiting for an EV to be moved out due to the maximum capacity limit. Moreover, we drop the C-rate of the EV battery from 1 C to 0.5 C in a variable charging rate scheme. For this reason, the EV will take a longer time to charge and stay in the smart parking lot for longer time intervals. However, our proposed LP-based simplified convex relaxation approach is found to outperforms the variable charging rate technique. The comparison is established based on the percentage of electricity cost-saving, the number of fully charged EV over the entire day, and average computational time. The aforementioned comparisons are tabulated for all DR events.

6.4. Comparative Analysis of Proposed IntelligentCharging Scheme with Variable Charging Rate Scheme

In Table 3, the power consumption cost comparison of the proposed algorithm is conducted with the variable charging rate scheme for different numbers of EVs and all DR programs.

Table 3. Comparison of Power Consumption Cost ($/day) and saving (in percentage).

	DR Event	Number of EV Arrived in a Day (*E*)				
		100	200	300	400	500
Variable Charging Rate	*No DR*	2694	5553	7079	7000	7245
	DR1	2597	5776	7081	7343	7385
	DR2	2661	5575	7495	7016	7153
	DR3	2687	6068	6888	6926	6960
Proposed	*No DR*	2600 (↓3.6%)	5457 (↑1.7%)	6655 (↓5.9%)	6668 (↓4.7%)	7028 (↓2.9%)
	DR1	2558 (↓1.5%)	5481 (↓5.1%)	6855 (↓3.1%)	7077 (↓3.6%)	7136 (↓3.3%)
	DR2	2564 (↓3.6%)	5476 (↓1.7%)	6903 (↓7.8%)	6928 (↓1.2%)	6800 (↓5.0%)
	DR3	2637 (↓1.8%)	5872 (↓3.2%)	6933 (↑0.6%)	6988 (↑0.8%)	6873 (↓1.2%)

The results make it evident that the proposed scheme is more cost-effective than other techniques. Moreover, In Table 4, we compared the number of EVs fully charged on a given day with a fixed number of charging pole ($X = 200$). Table 4 depicts the improved performance of the proposed algorithm compared to the variable charging rate scheme.

Table 4. Comparison of Number of Fully Charged EVs.

	DR Events	Number of EV Arrived in a Day (*E*)				
		100	**200**	**300**	**400**	**500**
Variable Charging Rate	*No DR*	100	200	203	203	210
	DR1	100	200	206	202	214
	DR2	100	200	201	207	205
	DR3	100	200	200	203	204
Proposed	*No DR*	100	200	203	215	216
	DR1	100	200	201	214	218
	DR2	100	200	201	213	219
	DR3	100	200	202	210	217

The performance of the proposed algorithm is also tested against the average SoC of the EVs at the departure time and tabulated in Table 5. The comparison is conducted for different numbers of vehicles $E = 100$ to 500. It is observed that the proposed technique performs better as the average SoC of an EV leaving the smart parking lot is 97% or above in five different vehicle counts. Moreover, the average time intervals required to achieve final SoC are also computed and compared in Table 6. Again, the proposed technique is clearing, which takes fewer time intervals to charge an EV on average. Therefore, the smart parking lot operator saves time to incorporate a greater number of EVs illustrated in Table 4, where a greater number of EV are charged during course of the entire day.

Table 5. Average State-of-Charge (SoC) of the EV's Battery at departure time.

E	**100**	**200**	**300**	**400**	**500**
Variable Charging Rate	0.99	0.99	0.94	0.93	0.93
Proposed	0.99	0.99	0.98	0.96	0.96

Table 6. Comparison for Average Time Intervals required to Achieve Final SoC.

E	**100**	**200**	**300**	**400**	**500**
Variable Charging Rate	23.02	23.10	20.19	21.12	20.56
Proposed	16.20	17.98	17.06	18.56	18.01

The computation time is also deduced for both algorithms on the same machine and listed in Table 7. It is found that the proposed algorithm marginally takes more computational time; however, the time difference is not increased to an alarming situation.

Table 7. Comparison of Average Computation Time (Seconds) for Algorithms under No DR Event.

E	**100**	**200**	**300**	**400**	**500**
Variable Charging Rate	5.18	5.21	5.37	5.37	5.40
Proposed	5.63	5.73	5.83	5.86	5.77

7. Conclusions and Future Work

A real-time and robust EV charging scheme is proposed for EV smart parking lots working under different DR programs. The objective of this proposed scheme is to charge the maximum number of EVs in a day at the minimum possible cost by taking into consideration DR events. The proposed LP and simplified convex relaxation-based on-off EV charging technique is also computationally viable for implementation in real-time scenarios. Moreover, the smart parking lot owner can participate in DR events and be recompensed by the utility after ensuring the variable or constant demand reduction. As soon as the demand limitation is committed for DR, the anticipated optimal EV charging scheduling strategy can start scheduling EVs for intelligent charging such that the overall demand for the load will remain within the constrained of prearranged demand limit. The benefits of the proposed charging scheme are two-fold: (1) the power consumption cost of the smart parking lot is minimized; and (2) the proposed EV intelligent charging technique has a minimal impact on the life cycle of the EV battery. The battery of the EV will always be charged at the maximum and fixed charging rate of its level-II charge standard and on C-rate equals to 1 C. The proposed intelligent charging technique will be tested in future under the presence of on-site renewable power generation sources. The variability of the available renewable power will be an interesting variable to deal in real-time. Finally, state-of-the-art machine learning techniques [36] are increasingly being deployed across a wide range of real-world optimization problems [37]. An interesting direction of future work could be to explore the incorporation of prediction models [38] to further enhance the capabilities of our proposed approach.

Author Contributions: Conceptualization, M.J. and M.B.Q.; Formal analysis, S.M.A.; N.S. and M.U.S.K.; Investigation, S.M.A.; N.S. and A.A.; Methodology, M.J. and M.B.Q.; Resources, M.J.; M.B.Q.; S.M.A.; N.S. and A.A.; Software, M.J. and M.B.Q.; Supervision, M.J. and R.N.; Validation, S.M.A.; N.S. and A.A.; writing–original draft preparation, M.J. and M.B.Q.; Writing–review and editing, M.J.; M.B.Q. and R.N. All authors have read and agreed to the published version of the manuscript.

Funding: This research received no external funding.

Acknowledgments: The authors like to thank Zeeshan Ullah for assisting with the writing and proofreading of the article.

Conflicts of Interest: The authors declare no conflict of interest.

References

1. Egbue, O.; Long, S. Barriers to widespread adoption of electric vehicles: An analysis of consumer attitudes and perceptions. *Energy Policy* **2012**, *48*, 717–729. [CrossRef]
2. Qian, K.; Zhou, C.; Allan, M.; Yuan, Y. Modeling of load demand due to EV battery charging in distribution systems. *IEEE Trans. Power Syst.* **2011**, *26*, 802–810. [CrossRef]
3. Qadir, H.; Khalid, O.; Khan, M.U.; Khan, A.U.R.; Nawaz, R. An optimal ride sharing recommendation framework for carpooling services. *IEEE Access* **2018**, *6*, 62296–62313. [CrossRef]
4. Shao, S.; Pipattanasomporn, M.; Rahman, S. Grid integration of electric vehicles and demand response with customer choice. *IEEE Trans. Smart Grid* **2012**, *3*, 543–550. [CrossRef]
5. Pipattanasomporn, M.; Kuzlu, M.; Rahman, S. An algorithm for intelligent home energy management and demand response analysis. *IEEE Trans. Smart Grid* **2012**, *3*, 2166–2173. [CrossRef]
6. Zhou, S.; Wu, Z.; Li, J.; Zhang, X.-P. Real-time energy control approach for smart home energy management system. *Elect. Power Compon. Syst.* **2014**, *42*, 315–326. [CrossRef]
7. Sajid, S.; Jawad, M.; Qureshi, M.B.; Khan, M.U.S.; Ali, S.M.; Khan, S.U. A Conditional-Constraint Optimization for Joint Energy Management of Data Center and Electric Vehicle Smart parking lot. In Proceedings of the Tenth International Green and Sustainable Computing Conference (IGSC), Alexandria, VA, USA, 21–24 October 2019.
8. Mehmood, F.; Khan, B.; Ali, S.M.; Qureshi, M.B.; Diver, C.; Nawaz, R. Multi-Renewable Energy Agent Based Control for Economic Dispatch and Frequency Regulation of Autonomous Renewable Grid. *IEEE Access* **2020**, *8*, 89534–89545. [CrossRef]
9. Su, W.; Chow, M.-Y. Computational intelligence-based energy management for a large-scale PHEV/PEV enabled municipal smart parking deck. *Appl. Energy* **2012**, *96*, 171–182. [CrossRef]
10. Rustam, M.A.; Khan, B.; Ali, S.M.; Qureshi, M.B.; Mehmood, C.A.; Khan, M.U.S.; Nawaz, R. An Adaptive Distributed Averaging Integral Control Scheme for Micro-Grids with Renewable Intermittency and Varying Operating Cost. *IEEE Access* **2019**, *8*, 455–464. [CrossRef]
11. Xu, Z.; Hu, Z.; Song, Y.; Zhao, W.; Zhang, Y. Coordination of PEV charging across multiple aggregators. *Appl. Energy* **2014**, *136*, 582–589. [CrossRef]
12. Huang, Q.; Jia, Q.S.; Qiu, Z.; Guan, X.; Deconinck, G. Matching EV charging load with uncertain wind: A simulation-based policy improvement approach. *IEEE Trans. Smart Grid* **2015**, *6*, 1425–1433. [CrossRef]

13. Kuran, M.S.; Viana, A.C.; Iannone, L.; Kofman, D.; Mermoud, G.; Vasseur, J.P. A smart parking lot management system for scheduling the recharging of electric vehicles. *IEEE Trans. Smart Grid* **2015**, *6*, 2942–2953. [CrossRef]
14. Shafie-khah, M.; Heydarian-Forushani, E.; Osório, G.J.; Gil, F.A.; Aghaei, J.; Barani, M.; Catalão, J.P. Optimal Behavior of Electric Vehicle Smart Parking Lots as Demand Response Aggregation Agents. *IEEE Trans. Smart Grid* **2016**, *7*, 2654–2665. [CrossRef]
15. He, Y.; Venkatesh, B.; Guan, L. Optimal scheduling for charging and discharging of electric vehicles. *IEEE Trans. Smart Grid* **2012**, *3*, 1095–1105. [CrossRef]
16. Jian, L.; Zhu, X.; Shao, Z.; Niu, S.; Chan, C.C. A scenario of vehicle-to-grid implementation and its double-layer optimal charging strategy for minimizing load variance within regional smart grids. *Energy Convers. Manag.* **2014**, *78*, 508–517. [CrossRef]
17. Zhang, L.; Li, Y. A game-theoretic approach to optimal scheduling of smart parking lot electric vehicle charging. *IEEE Trans. Veh. Technol.* **2016**, *65*, 4068–4078. [CrossRef]
18. Ansari, M.; Al-Awami, A.T.; Sortomme, E.; Abido, M.A. Coordinated bidding of ancillary services for vehicle-to-grid using fuzzy optimization. *IEEE Trans. Smart Grid* **2015**, *6*, 261–270. [CrossRef]
19. Jin, C.; Tang, J.; Ghosh, P. Optimizing electric vehicle charging with energy storage in the electricity market. *IEEE Trans. Smart Grid* **2013**, *4*, 311–320. [CrossRef]
20. Han, S.; Han, S.; Sezaki, K. Development of an optimal vehicle-to-grid aggregator for frequency regulation. *IEEE Trans. Smart Grid* **2010**, *1*, 65–72.
21. Shafie-khah, M.; Heydarian-Forushani, E.; Golshan, M.E.; Siano, P.; Moghaddam, M.P.; El-Eslami, M.K.; Catalão, J.P. Optimal trading of plug-in electric vehicle aggregation agents in a market environment for sustainability. *Appl. Energy* **2016**, *162*, 601–612. [CrossRef]
22. Yazdani-Damavandi, M.; Moghaddam, M.P.; Haghifam, M.; Shafie-khah, M.; Catalão, J.P. Modeling Operational Behavior of Plug-in Electric Vehicles' Smart parking Lot in Multienergy Systems. *IEEE Trans. Smart Grid* **2016**, *7*, 124–135. [CrossRef]
23. Škugor, B.; Deur, J. Dynamic programming-based optimisation of charging an electric vehicle fleet system represented by an aggregate battery model. *Energy* **2015**, *92*, 456–465. [CrossRef]
24. Lausenhammer, W.; Engel, D.; Green, R. Utilizing capabilities of plug in electric vehicles with a new demand response optimization software framework: Okeanos. *Int. J. Electr. Power Energy Syst.* **2016**, *75*, 1–7. [CrossRef]
25. Tan, Z.; Yang, P.; Nehorai, A. An optimal and distributed demand response strategy with electric vehicles in the smart grid. *IEEE Trans. Smart Grid* **2014**, *5*, 861–869. [CrossRef]
26. Rassaei, F.; Soh, W.-S.; Chua, K.-C. Demand response for residential electric vehicles with random usage patterns in smart grids. *IEEE Trans. Sustain. Energy* **2015**, *6*, 1367–1376. [CrossRef]
27. Yao, L.; Damiran, Z.; Lim, W.H. Optimal Charging and Discharging Scheduling for Electric Vehicles in a Smart parking Station with Photovoltaic System and Energy Storage System. *Energies* **2017**, *10*, 550. [CrossRef]
28. Cao, Y.; Kaiwartya, O.; Wang, R.; Jiang, T.; Cao, Y.; Aslam, N.; Sexton, G. Toward Efficient, Scalable, and Coordinated On-the-Move EV Charging Management. *IEEE Wirel. Commun.* **2017**, *24*, 66–73. [CrossRef]
29. Cao, Y.; Jiang, T.; Kaiwartya, O.; Sun, H.; Zhou, H.; Wang, R. Toward Pre-Empted EV Charging Recommendation Through V2V-Based Reservation System. *IEEE Trans. Syst. Man. Cybern. Syst.* **2019**, 1–14. [CrossRef]
30. ERCOT. Meteorological Data of Texas from National Estuarine Research Reserve System [Online]. ERCOT. Available online: http://cdmo.baruch.sc.edu/get/export.cfm/ (accessed on 23 April 2020).
31. Jawad, M.; Qureshi, M.B.; Khan, U.; Ali, S.M.; Mehmood, A.; Khan, B.; Khan, S.U. A robust Optimization Technique for Energy Cost Minimization of Cloud Data Centers. *IEEE Trans. Cloud Comput.* **2018**, 1. [CrossRef]
32. Şengör, İ.; Erdinç, O.; Yener, B.; Taşcıkaraoğlu, A.; Catalão, J.P. Optimal energy management of EV smart parking lots under peak load reduction-based DR programs considering uncertainty. *IEEE Trans. Sustain. Energy* **2019**, *10*, 1034–1043. [CrossRef]
33. Jawad, M.; Qureshi, M.B.; Nadeem, A.; Ali, S.M.; Shabbir, N.; Rafiq, M.N. Bi-Directional Nano Grid Design for Organizations with Plug-In Electric Vehicle Charging at Workplace. In Proceedings of the 2018 IEEE International Conference on Electro./Information Technology (EIT), Rochester, MI, USA, 3–5 May 2018; pp. 357–361.

34. Nadeem, A.; Rafiq, M.N.; Qureshi, M.B.; Jawad, M. Joint Power Management of Telecom Exchanges and Electric Vehicles Using Hybrid AC-DC Microgrid. In Proceedings of the International Conference on Frontiers of Information Technology (FIT), Islamabad, Pakistan, 18–20 December 2017; pp. 127–132.
35. Mao, T.; Zhang, X.; Zhou, B. Intelligent Energy Management Algorithms for EV-charging Scheduling with Consideration of Multiple EV Charging Modes. *Energies* **2019**, *12*, 265. [CrossRef]
36. Yunus, R.; Arif, O.; Afzal, H.; Amjad, M.F.; Abbas, H.; Bokhari, H.N.; Haider, S.T.; Zafar, N.; Nawaz, R. A Framework to Estimate the Nutritional Value of Food in Real Time Using Deep Learning Techniques. *IEEE Access* **2019**, *7*, 2643–2652. [CrossRef]
37. Anwaar, F.; Iltaf, N.; Afzal, H.; Nawaz, R. HRS-CE: A hybrid framework to integrate content embeddings in recommender systems for cold start items. *J. Comput. Sci.* **2018**, *29*, 9–18. [CrossRef]
38. Ayyaz, S.; Qamar, U.; Nawaz, R. HCF-CRS: A Hybrid content based fuzzy conformal recommender system for providing recommendations with confidence. *PLoS ONE* **2018**, *13*, e0204849. [CrossRef] [PubMed]

sensors

MDPI

Article

A Novel Framework Using Deep Auto-Encoders Based Linear Model for Data Classification

Ahmad M. Karim [1], Hilal Kaya [1], Mehmet Serdar Güzel [2], Mehmet R. Tolun [3], Fatih V. Çelebi [1] and Alok Mishra [4,5,*]

1 Computer Engineering Department, AYBU, Ankara 06830, Turkey; ahmad.mozaffer.karim@gmail.com (A.M.K.); hilalkaya@ybu.edu.tr (H.K.); fatihvcelebi@gmail.com (F.V.Ç.)
2 Computer Engineering Department, Ankara University, Ankara 06830, Turkey; mguzel@ankara.edu.tr
3 Computer Engineering Department, Konya Food and Agriculture University, Konya 42080, Turkey; mehmet.tolum@gidatarim.edu.tr
4 Faculty of Logistics, Molde University College-Specialized University in Logistics, 6402 Molde, Norway
5 Software Engineering Department, Atilim University, Ankara 06830, Turkey
* Correspondence: alok.mishra@himolde.no

Received: 27 September 2020; Accepted: 5 November 2020; Published: 9 November 2020

check for updates

Abstract: This paper proposes a novel data classification framework, combining sparse auto-encoders (SAEs) and a post-processing system consisting of a linear system model relying on Particle Swarm Optimization (PSO) algorithm. All the sensitive and high-level features are extracted by using the first auto-encoder which is wired to the second auto-encoder, followed by a Softmax function layer to classify the extracted features obtained from the second layer. The two auto-encoders and the Softmax classifier are stacked in order to be trained in a supervised approach using the well-known backpropagation algorithm to enhance the performance of the neural network. Afterwards, the linear model transforms the calculated output of the deep stacked sparse auto-encoder to a value close to the anticipated output. This simple transformation increases the overall data classification performance of the stacked sparse auto-encoder architecture. The PSO algorithm allows the estimation of the parameters of the linear model in a metaheuristic policy. The proposed framework is validated by using three public datasets, which present promising results when compared with the current literature. Furthermore, the framework can be applied to any data classification problem by considering minor updates such as altering some parameters including input features, hidden neurons and output classes.

Keywords: deep sparse auto-encoders; medical diagnosis; linear model; data classification; PSO algorithm

---

1. Introduction

Deep learning (DL) is a new paradigm of neural networks, which is employed in different fields such as image classification and recognition, medical imaging and robotics etc. The deep auto-encoder (DAE) is also a popular deep learning technique and has been recently adapted to various applications in different fields [1–4]. Bhatkoti and Paul propose a new framework for Alzheimer's disease diagnosis based on deep learning and the KSA algorithm. In this application, the results of the modified approach are compared to the non-modified k-sparse method. The σKSA algorithm optimizes the competence of diagnosis compared to the previous research [5,6]. Tong et al. present a software defect prediction application by using the advantages of stacked denoising auto-encoders (SDAEs) and a two-stage ensemble (TSE). In the first step, SDAEs are used to learn the deep representations from the imitative software metrics. Moreover, a new ensemble learning method, TSE, is proposed to predict the label imbalance problem. The proposed method is trained and tested by using 12 NASA benchmark test data to show the effectiveness of the SDAEsTSE system, which is significantly effective for software defect prediction [7].

Kuo et al. (2017) propose a stacked denoising auto-encoder for building a deep network for student dropout prediction. The system is trained with recent years' data and is used to estimate the results of the current year for counseling in order to warn of students who might drop out [8].

Another leading study trains an auto-encoder neural network to encode and decode a geochemical data with unidentified composite multivariate possibility distributions. During the training, rare event examples contribute to the deep auto-encoder network. These examples can be classified by the trained network as abnormal examples due to their reasonably greater reconstructed mistakes. The Southwestern Fujian district in China is selected as a case research field [8].

Han et al. present some ideas of a deep sparse auto-encoder mixed with compressed sensing (CS) theory, which can enhance the compacted selection process of CS with compressing of the sparse auto-encoder in deep learning. The innovative CS theory does not provide any function of autonomic instruction, so they present the notion of a stacked auto-encoder of a deep neural network to optimize the theory. At that point, they compute the mistakes between the retrieval of the input and output features. By adjudicating the achieved error and the suitable error, the stacked auto-encoder compressed sensing model can select separately the best suitable sparsity and the best suitable length of dimension vector [9]. Salaken et al. propose a deep auto-encoder classification technique which primarily learns high-level features and then trains an artificial neural network (ANN) by employing these learned features. Experimental results prove that the technique offers satisfactory results when compared with other state-of-the arts classifiers when trained with the same features and the training set [10]. Khatab et al. present a novel technique which takes advantage of deep learning and deep extracted features by employing an auto-encoder to enhance the localization achievement in the feature learning and the classification. Moreover, the fingerprint dataset also needs to be reorganized, so the authors increase the training data number, so as to enhance the localization achievement, progressively. Experiments show that the presented technique supplies an important enhancement in localization achievement by using deep features extracted by an auto-encoder and increasing the training data number [11].

In addition to the deep auto-encoder neural network, a convolutional neural network has effective applications. Khan et al. offer a new convolutional neural network and random forest estimator to categorize the complex time series input, identifying whether it agrees with a breathing activity. Furthermore, the authors collect a comprehensive dataset for training the proposed method and evolve reference benchmarks for future studies comprising the field. According to the obtained results, they conclude that convolutional neural networks mixed with passive radars show high potential for a taxonomy of human actions [12]. Tang et al. propose a novel method, involving a preprocessing step, supported with two deep auto-encoders. Within the pre-processing stage, the input data are divided into segments, and then formal information is extracted so as to feed auto-encoders. It is claimed that this method produces acceptable results when compared with CNN-based feature learning approaches [13]. Yin et al. propose a new approach to explore an intrusion recognition system depending on a deep neural network. They propose a model for "intrusion recognition" based on recurrent neural networks ("RNN-IDS"). Furthermore, the system can achieve classification process as both a binary and multiclass classifier. The proposed model is compared with random forest, J48, SVM, ANN, and other machine learning techniques presented in earlier studies on the commonly used dataset. The experiments prove that "RNN-IDS" is actually appropriate for demonstrating a reliable classification model and outperforms well-known machine learning classification methods in binary and multiclass classification problems [14]. Yu et al. propose a technique to automatically classify the fetal facial standard plane (FFSP) by using a deep convolutional neural network (DCNN) method. The technique involves "16" convolutional layers, having small size "3×3" kernels, and also fully-connected layers (FCLs) layers. To reduce DCNN parameters, a "global average pooling" is adopted into the last pooling layer, which relieves the overfitting status and mends the achievement under fixed training data. The transfer learning technique followed by a data increase method, appropriate for FFSP, are executed to increase the classification accuracy gradually. Comprehensive experiments validate the benefit of the proposed approach through classical methods and the performance of DCNN to classify fetal facial standard plane for clinical detection [15].

Visual surveying of the large size of data has drawbacks and weaknesses. Visual investigation is time-consuming and may encounter conflicts in recognition, classification and detection processes, which are fundamental problems of large size of data. Therefore, many computer-aided diagnosis systems are proposed for data classification and processing by using machine learning techniques etc. Despite the researchers' recent interest, it is still an open field and needs further solutions. This essentially motivates authors to contribute in this field. Accordingly, as aforementioned, this study introduces a general framework for data classification and processing issues. This framework is verified by employing a number of benchmark datasets in different fields. Overall, the main advantage of this framework is its remarkable experimental results when compared with former studies. Furthermore, the proposed framework can be used in any field with minimum effort by setting the model parameters based on the characteristics of the problem.

Generally, the two sparse auto-encoders are utilized to diminish the dimension of input features and learn refined features. Those features are then classified by employing a Softmax layer. The whole model is stacked to provide a supervised training methodology. Then, the critical contribution is achieved by integrating a linear model, utilizing a metaheuristic algorithm for optimization, and is applied to enhance the deep sparse auto-encoder performance.

2. Literature Review

Several studies related to three different datasets (epileptic seizure detection, cardiac arrhythmia and SPECTF classification) are analyzed and presented in Tables 10–12. Epileptic seizure is one of the most studied diseases in the field of computer-aided detection systems. Srinivasan et al. propose a new system based on time–frequency domain for feature extraction, and RNN were used to classify the features. The proposed method presents 99.60% accuracy as can be seen in [16]. Subasi and Ercelebi propose artificial neural network (ANN)-based wavelet transform (WT) and produce only 92% performance [17]. Subasi proposes a discrete WT based on a mixture of expert model, which presents only 94.5% performance [18]. Kannathal et al. propose a dynamic neuro-fuzzy inference system (ANFIS) based on entropy measures and produce 95% performance as can be seen in [19]. Tzallas et al. propose a new method based on time–frequency analysis and ANN which produces a high accuracy of 100% [20]. Polat and Güneş propose a fast Fourier transformation and decision tree (DT) which presents 98.72% performance [21]. Acharya et al. employ wavelet packet decomposition (WPD) to decompose segments and principal component analysis (PCA) to extract eigenvalues from the coefficients. Then, a supervised technique, namely, Gaussian mixture model (GMM) classifier, is employed to categorize the extracted features and obtain 99% accuracy [22]. Acharya et al. propose a combination of entropies, "HOS", "Higuchi FD", "Hurst exponent" and FC, and the proposed method offers "99.70%" accuracy [23]. Peker et al. propose a complex-value artificial neural network (CVANN) based on dual tree complex wavelet transformation (DTCWT). The proposed method presents 100% performance [24]. Karim et al. propose a new framework involving deep sparse auto-encoders (DSAE) utilizing the Taguchi optimization method, and the proposed method presents 100% accuracy [25]. Recently, Karim et al. modified the same framework by incorporating energy spectral density function, used to extract features, into a similar DSAE architecture. The results reveal that it outperforms many existing systems, especially in medical datasets [26].

Additionally, an important study in arrhythmias relying on spontaneous methods was recently offered, in which a model for estimation of cardiac arrhythmias is proposed [27]. The presented method applies two conventional supervised techniques (k-NN and SVM), respectively. The proposed method is validated and tested by employing the "UCI" dataset. While k-NN presents "73.8%" accuracy rate, SVM surprisingly achieves a 68.8% accuracy rate. Mustaqeem et al. propose a novel system for the recognition of arrhythmia, according to which, a wrapper algorithm is initially used to select effective features from the UCI dataset. Then, different classifiers, namely MLP, KNN, SVM,

RFT and NB are combined with the proposed feature-extracted algorithm, respectively. The validation accuracies demonstrate that the MLP achieves a suitable result of 78.26%, whereas the results obtained for SVM and KNN are 74.4% and 76.6%, respectively [28]. Zuo et al. present a technique for the taxonomy of cardiac arrhythmia using a k-nearest neighbor classifier. The submitted method outperforms traditional KNN algorithms and produces more than 70% accuracy [29]. Besides that, an ANN-based architecture is applied to classify the Electrocardiography (ECG) records for cardiac arrhythmia taxonomy. It is claimed that the experimental results yield more than 87% classification accuracy [30]. Moreover, Persada et al. propose Best First and CsfSubsetEval for the feature selection process. The selected features are classified by using several classifiers, and the best precision is obtained by using the "RBF Classifier" in the combination of BFS and "CsfSubsetEval" techniques, producing 81% [31]. Jadhav et al. propose a modular neural network model for the binary classification (normal or abnormal) of arrhythmia dataset. The proposed model is claimed to attain 82.22% accuracy with the given dataset [32]. Further corresponding studies can be found in [33–35].

Moreover, a number of previous studies in the field of SPECTF classification are accessible. Srinivas et al. propose an SVM technique relying on sparsity-based dictionary learning. The proposed method presents 97.8% accuracy [36]. An alternative study offers a Bayesian network to select features. The method entails a vast number of features and produces 95.76% accuracy [37]. Cha et al. propose a new data description approach, namely support vector data description, which is assessed by employing datasets from the UCI repository. The method achieves almost 95% accuracy for the given dataset [38]. Furthermore, Liu et al. propose a new SVDD-based method. The proposed method offers 90% accuracy [39]. Previously, Cui et al. combined an improved version of k-nearest neighbors and the method is known as transductive confidence machine (TCM). The authors claim that this approach (TCM-IKNN) presents 90% accuracy with the UCI dataset [40]. Alternatively, a previous study on discretization approach, namely, "core-generating approximate minimum entropy discretization", was also presented by [41]. This aims to control the lowermost entropy cuts in order to create discrete data points providing nonempty cores. The presented method is also confirmed by employing the UCI dataset and achieves 84% accuracy rate [41].

3. Material and Methods

The main contribution of this paper is to integrate a post processing procedure to a data classification framework. Accordingly, a strong deep learning framework combining sparse auto-encoders (SAEs) followed by a Softmax Classifier, a generalization of the binary form of the Logistic Regression method, is initially designed. The auto-encoder levels and the classifier level are stacked so as to be trained in a supervised approach based on a backpropagation algorithm. In order to increase the overall classification accuracy, a linear transformation function is integrated into the framework. This layer, in essence, improves the results obtained from DAEs based on a linear model. The critical issue here is to estimate the optimum parameter for the linear transformation model. A strong and reliable metaheuristic algorithm, PSO, is employed to approximate the most optimum model parameters. All these steps are detailed in the following sub sections.

3.1. Stacked Sparse Auto-Encoder

The stacked sparse auto-encoder (SSAE) is principally a neural network involving of a number of auto-encoders where each auto-encoder represents a layer and is trained in an unsupervised fashion using unlabeled data. The input of each auto-encoder is the output of the previous one. The training of an auto-encoder estimates the optimal parameters by using different algorithms which reduce the divergence between input x and output $\dot{x}$. The coding between input and output is represented by the equations illustrated below. Here, the input vector $x = (1, 2, 3, 4 \ldots, N)$, is transformed into hidden representation "$\dot{x}$", by employing a nonlinear model.

$$\dot{x} = f(x) = M_f(W_1 x + b_1) \quad (1)$$

$$n_1^{(1)} = M_f(w_{11}^{(1)} x_1 + \cdots w_{15}^{(1)} x_5 + b_1^{(1)}) \tag{2}$$

$$n_i^{(1)} = M_f(w_{i1}^{(1)} x_1 + \cdots w_{i5}^{(1)} x_5 + b_i^{(1)}) \tag{3}$$

Here $n_i^{(1)}$ refers to the ith neuron at the first layer for the architecture, M is an activation function, $w_{i,}$ and b_i refer to weight matrix and the bias parameter, respectively.

The final mathematical model is illustrated in Equation (4):

$$n_{w,b}(x) = M_f(w_{11}^{(2)} n_1^{(2)} + \cdots w_{15}^{2} n_5 + \cdots + b_1^{(2)}) \tag{4}$$

The input x and output $\dot{x}$ discrepancy is represented by using a cost function. Several algorithms are used to find the optimum parameters of the network. The corresponding mathematical model can be seen in [25,42]. The model of Stacked Sparse Auto-encoder (SSAE), used in the proposed framework, is illustrated in Figure A1 and can be seen in Appendix A. The model has two hidden layers and a classifier layer (SoftMax).

3.2. The Particle Swarm Optimization (PSO) Algorithm

PSO algorithms are considered as population-based metaheuristic algorithms proposed by [43–46]. These algorithms impersonate the social behavior of birds for problem solving.The PSO algorithm is set with a group of arbitrary solutions, representing the particles, and then it explores to approximate an optimal solution by updating the generations. In each iteration, every particle is modified by considering the two (best) values, namely local and global best values. The first best solution that is attained so far by the particle itself is denoted as the best local solution and is stored, known as "pbest" value. Then, the other, global, refers the best solution achieved thus far by a particle located in the population, and this best solution is a global best, known as "gbest" value. The particle updates the positions and velocity by employing Equations (5) and (6) after selecting the best two solutions.

$$X_{k+1}^i = X_k^i + V_{k+1}^i \tag{5}$$

$$V_{k+1}^i = wV_k^i + c_1 r_1 \left(P_k^i - X_k^i\right) + c_2 r_2 \left(P_k^g - X_k^i\right) \tag{6}$$

Here, X_k^i represents particle position, V_k^i represents particle velocity, P_k^i represents the best "remembered" individual particle position (pbest), P_k^g represents the best swarm position (gbest), c_1 and c_2. are cognitive and social parameters. Additionally, r_1, r_2 are random parameters between (0,1) and w refers inertial coefficient (0,1). This manipulates convergence and "explore-exploit" trade-off in the PSO algorithm. PSO algorithms offers a number of advantages when compared with other optimization algorithms. PSO is a fast optimization algorithm and only needs few parameters for tuning. Especially, when PSO is compared with one of its main counterpart algorithms, Genetics Algorithm (GA), it should be noted that PSO can converge faster and needs fewer parameters to be configured.

Accordingly, PSO is successfully applied in several fields, such as neural networks, optimization problems, etc. Algorithm 1 refers to the conventional PSO algorithm [47].

Algorithm 1. Pseudo Code of PSO Algorithm.

```
        For each particle
          Set particles in a random manner
        End
        Do
            Estimate the Local best "pBest" for each particle
            If the "pBest" is enhanced
          Update "pBest" value
        End
          Global Best (gBest) is updated as the best of "pBests"
          For each particle
             Estimate the velocity of particles via Equations (5) and (6)
             Update the positions of the particles
           End
End
```

3.3. A New Deep Learning Framework Using Deep Auto-Encoders and a Linear Model Based on PSO

Suppose a trained deep stacked auto-encoder is used to classify an object into one of the "M" classes. The input layer of the deep stacked auto-encoder involves "N" neurons that are related to object features $X_1, X_2, \ldots, X_N$, and the output layer involves "M" neurons that stand for the expected output (class label) $\hat{Z}_1, \hat{Z}_2, \hat{Z}_3, \hat{Z}_M$ (see Figure 1).

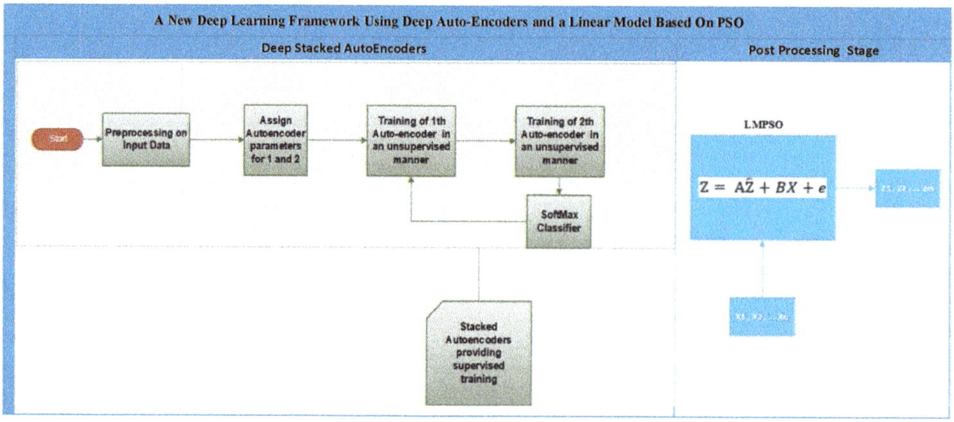

Figure 1. The Deep Learning Framework Based on a Linear Model and metaheuristic algorithm (PSO).

The deep auto-encoder involves two auto-encoders and Softmax, where the auto-encoders try to learn the high-level features from the input data X. The aim of using a number of auto-encoders is to reduce the number of features gradually. This is because dropping the number of features suddenly in one auto-encoder can lead to missing important features and affect the accuracy. The cost function of the stacked auto-encoders is represented as Equation (7).

$$E = \frac{1}{N}\sum_{n=1}^{N}\sum_{k=1}^{K}(x_{kn} - \hat{x}_{kn})^2 + \lambda * \boldsymbol{\Omega}_{weights} + \beta * \boldsymbol{\Omega}_{sparsity} \quad (7)$$

Here, the error rate is denoted by E, the input features are illustrated by "x", the reconstructed features are illustrated with "$\hat{x}$", λ is the coefficient for the "L2 Weight Regularization", β is the coefficient for "Sparsity Regularization", and $\boldsymbol{\Omega}_{weights}$ signifies the "L2 Weight Regularization", which can be represented as shown in Equation (8).

$$\boldsymbol{\Omega}_{weights} = \frac{1}{2}\sum_{l}^{L}\sum_{j}^{n}\sum_{i}^{k} w_{ji}^{(l)^2} \tag{8}$$

Here, L presents the number of hidden layers, n is for the number of observations, and k indicates the variable number of the current training data.

Finally, $\boldsymbol{\Omega}_{sparsity}$ is the Sparsity Regularization parameter which adjusts the degree of sparsity of the output from the hidden layers, as illustrated in Equation (9).

$$\boldsymbol{\Omega}_{sparsity} = \sum_{i=1}^{D(1)} KL(\rho \| \hat{\rho}_i) = \sum_{i=1}^{D(1)} \rho \, log(\rho \| \hat{\rho}_i) + (1-\rho) log\left(\frac{1-\rho}{1-\hat{\rho}_i}\right) \tag{9}$$

Here, the desired value is represented by ρ, $\hat{\rho}_i$ symbolizes the average output activation of any neuron i, and KL represents the function, measuring the variation between two probability distributions based on the same data. Furthermore, the features that produce minimum cost in Equation (1) are selected and become input to Softmax, see Equation (10). Softmax is exploited as a classifier of the extracted features from X to the labels Z (see Figure 1).

$$Q_{SoftMax}\left(z^i\right) = \frac{e^{z^{(i)}}}{\sum_{j=0}^{k} e^{z_k^{(i)}}} \tag{10}$$

Here the net input z is defined as

$$z = \sum_{l=0}^{m} w_l \, x_l \tag{11}$$

Here, while w represents the weight vector, x symbolizes the feature vector of lth training sample. Essentially, the Softmax function calculates the probability of belonging to a class "j" for a training sample "$x^{(i)}$" by taking into account the given weights and net input $z^{(i)}$. Softmax is used without other classifiers because it is a transfer function and multiclass classifier which acts like an output layer to the previous auto-encoders. Then, the auto-encoders and Softmax layers are combined and trained by using a backpropagation algorithm in a supervised fashion to improve the performance of the network.

Moreover, antithetically to previous deep learning applications, the output of the deep auto-encoder does not generate the final prediction but optimizes it by using a linear model [48]. Essentially, the performance of a deep networks is considered by the network's structure, transfer function, and learning algorithm. Yet, a network classifier tends to be weak once it is designed based on an inappropriate structure. Essentially, there is no certain way to estimate a proper structure. A recent study proposed a linear model as a post processing layer based on Kalman Filter to improve overall classification performance [49]. Our study is inspired by this previous work and it employs the linear model so as to transform the predicted output of the network to a value close to the desired output via the linear combination of the object features and the expected output. This simple transformation can be considered as a post processing step, reducing the error of network and enhancing classification performance. A metaheuristic approach, PSO, is employed to optimize the parameters of the linear model. Overall, the parameters of the Linear model are calculated during the iteration of PSO algorithm. The linear model utilizes the predicted output of the deep network and the object features as input to estimate the class labels. The output of the DSAE $\hat{Z}$ is processed in a linear model by using X, coefficients A, B and the error rate e to produce the optimized result Z (see Equation (12)).

$$Z = A\hat{Z} + BX + e \tag{12}$$

Here, A represents diagonal matrix $M \times M$ as shown in (13), B denotes $M \times N$ matrix as shown in (14), and e is for the error rate. Moreover, coefficients, namely A and B, are unknown for the linear model [50]. The values of A and B are estimated by using a PSO algorithm, and the parameters of PSO are selected depending on the problem type and input features.

$$A = diag\ [a_{11}\quad a_{22}\quad a_{MM}] \tag{13}$$

$$B = \begin{bmatrix} b_{11} & \cdots & b_{1N} \\ \vdots & \ddots & \vdots \\ b_{M1} & \cdots & b_{MN} \end{bmatrix} \tag{14}$$

The details of the linear model mathematics are explained in [49], and the whole framework flowchart is illustrated in Figure 2.

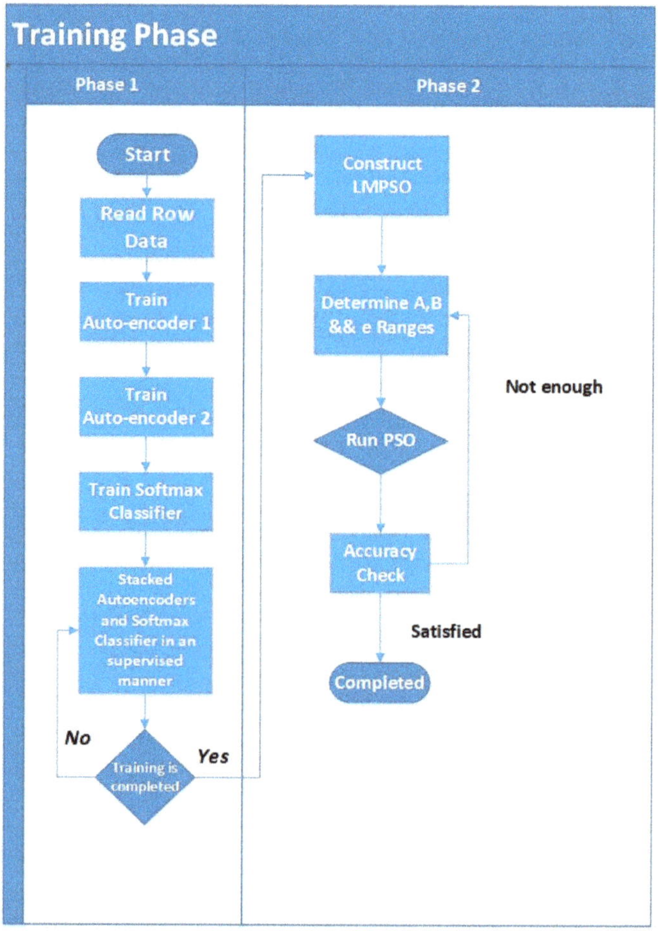

Figure 2. Training Flowchart for the Proposed Framework.

In each iteration of PSO, the predicted Z is controlled by using MSE with optimal prediction Q, as illustrated in Equation (15).

$$MSE = \frac{1}{m}\sum_{i=1}^{m}(Q_i - Z_i)^2 \quad (15)$$

Here, m denotes the number of examples, Q_i is the optimum class label for input features and MSE is the discrepancy rate between the z_i and Q_i.

The MSE is represented as a cost function. PSO minimizes its value by estimating the best values for parameters A, B and e.

4. Experimental Results

The parameters calculated to improve the performance of the proposed framework are: "True Positive Rate" (Recall), "True Negative Rate" (TNR), "positive predictive value" (Precision), "negative predictive value" (NPV), "false positive rate" (FPR), "false discovery rate" (FDR), "miss rate" (MR), "accuracy" (ACC), "F1 score" (F1-s) and "Matthews correlation coefficient" (MCC). Definitions of these parameters can be seen authors' previous work [25].

Each dataset has been divided into test and training sets according to the preliminary experiments and based on our previous studies. According to these, the Epileptic Seizure dataset is divided as 100 samples for training and the other 100 for testing, indicating 50% for test and 50% for training. The SPECTF Classification dataset, on the other hand, is arranged as 187 (70%) for training, and 87 (30%) for the testing process. The final dataset, the cardiac arrhythmias dataset, consists of 450 instances from 16 classes with 70% of those data employed for training and 30% for the testing procedures, respectively. Overfitting is a critical problem for classification models. In order to prevent overfitting, a random subsampling validation technique was applied during the training process. Following this, each experiment is repeated five times and the average of those experiments is registered.

4.1. Epileptic Seizure

The proposed framework is confirmed by employing a popular public dataset provided by Bonn University [51]. The dataset consists of 200 samples, with each sample consisting of 4096 features. The EEG data is split into two groups for training and testing procedures. Each group involves 100 examples, 50 of which are normal and the remaining 50 are abnormal. Those cases are illustrated in Figure 3. According to the framework, the first and second auto-encoders extract high-level features obtained from EEG signals and then diminish the number of features to 2007 and 112, respectively. Details of the parametric configuration of auto-encoders are shown in Table 1. Later, the Softmax layer classifies the extracted features as being normal and abnormal.

The linear model is then used to enhance the results, and the parameters of the linear model are estimated by using the PSO algorithm. The linear model parameters are estimated in 30 epochs and a reasonable MSE value is produced, as shown in Figure 4. Besides, the parameters of PSO are presented in Table 2.

Table 1. Auto-Encoder Parameters for Epileptic Seizure Detection.

Parameter	First Auto-Encoder	Second Auto-Encoder
Hidden Layer Size (HLS)	2007	112
Max Epoch Number (MEN)	420	110
L2 Regularization Parameter	0.004	0.002
Sparsity Regularization (SR)	4	2
Sparsity Proportion (SP)	0.14	0.12

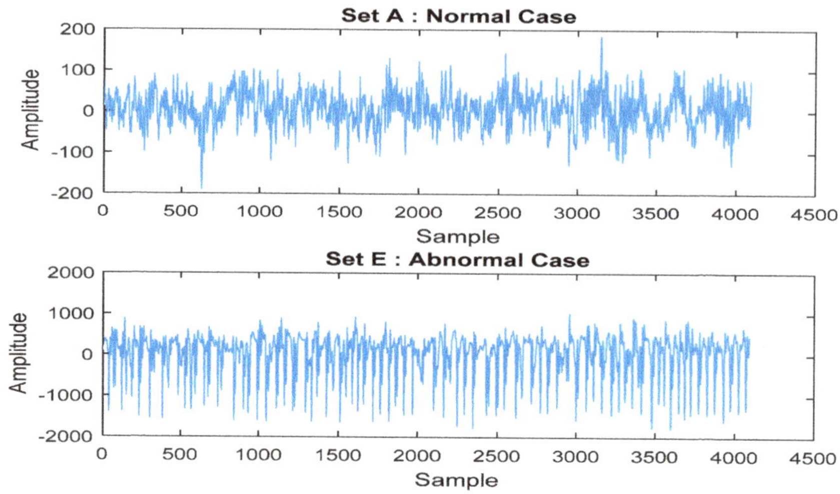

Figure 3. Datasets for Normal and Abnormal Cases.

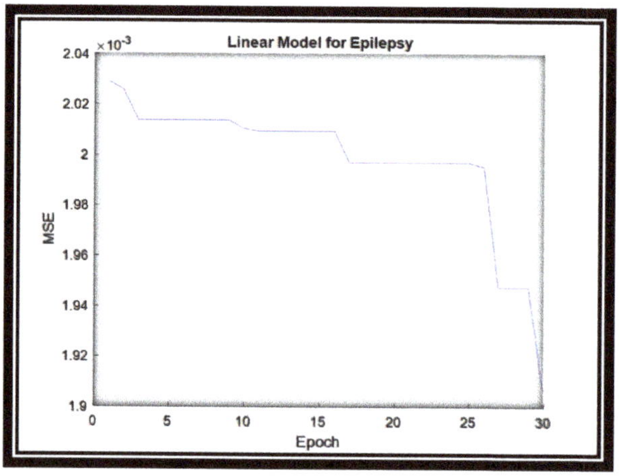

Figure 4. The MSE for the Linear System for Epilepsy dataset.

Table 2. PSO Parameters for Epileptic Seizure Detection.

PSO Parameter	Value
Number of particles	50
Maximum iteration	30
Cognitive parameter	2
Social parameter	2
Min inertia weight	0.9
Max inertia weight	0.2

The test process is repeated five times with the same parameters and hidden layer values, but in each implementation the training and test data are arbitrarily designated to avoid overfitting. The average results of the dataset based on previously defined evaluation parameters is shown in Table 3. The corresponding table represents the results during the testing process.

Table 3. Epileptic Seizure Detection Results.

Parameter	DSAEs without Post-Processing	DSAEs Using PSO
Recall	0.9348	1.0000
TNR	0.7222	1.0000
Precision	0.7414	1.0000
NPV	0.9286	1.0000
ACC	0.8200	1.0000
F1-s	0.8269	1.0000
MCC	0.6634	1.0000

4.2. SPECTF Classification

The proposed framework is assessed by employing another benchmark dataset, namely, "SPECTF, (Single Proton Emission Computed Tomography) Heart datasets", which is mainly presented in [52]. This dataset involves "normal" and "abnormal" classes that comprise more than 267 examples, with each of these instances consisting of 44 features. There exists 40 occurrences of each class at the training dataset, whereas the validation dataset contains "172 normal" and "15 abnormal" examples. As it is noted, auto-encoders can reduce the input dimension, and accordingly, the features in auto-encoders 1 and 2 are reduced step-by-step to 40 and 35, respectively, which essentially extracts high-level and sensitive features from input data.

The constraints of the auto-encoders are illustrated in Table 4. The parameters of the PSO algorithm are presented in Table 5.

Table 4. Auto-Encoder Parameters for Single Proton Emission Computed Tomography (SPECTF) Classification.

Parameter	Auto-Encoder 1	Auto-Encoder 2
Hidden Layer Size (HLS)	40	35
Max Epoch Number (MEN)	110	60
L2 Regularization Parameter	0.003	0.001
Sparsity Regularization (SR)	2	1
Sparsity Proportion (SP)	0.1	0.1

Table 5. PSO Parameters for SPECTF Classification.

PSO Parameter	Value
Number of particles	40
Maximum iteration	40
Cognitive parameter	2
Social parameter	2
Min inertia weight	0.9
Max inertia weight	0.2

The experimental results are evaluated by calculating the values of parameters, as presented in Table 6.

Table 6. SPECTF Classification Results.

Parameter.	DSAEs without Post-Processing	DSAEs Using PSO
Recall	*0.9554*	*1.0000*
TNR	*0.3333*	*0.8750*
Precision	*0.8824*	*0.9884*
NPV	*0.5882*	*1.0000*
ACC	*0.8556*	*0.9893*
F1-s	*0.9174*	*0.9942*
MCC	*0.3686*	*0.9300*

For this dataset, the linear model parameters are converged in almost 20 epochs and produce "2.03" error value as illustrated in Figure 5.

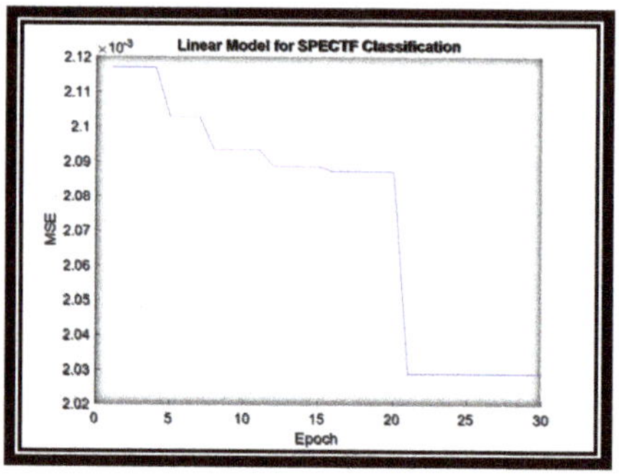

Figure 5. The MSE for the Linear System for SPECTF dataset.

4.3. *Diagnosis of Cardiac Arrhythmia*

The final benchmark dataset involves the data regarding cardiac arrhythmia, presented in [52]. This dataset consists of 450 instances from 16 different classes. Each class has 279 features. The proposed framework is trained for this dataset, according to which, the first Auto-Encoder is trained by employing an unsupervised approach and achieves a decrease in the number of features from 279 to 250. The output of the first one is passed to the second auto-encoder, which is also trained in an unsupervised manner. Afterwards, the number of features is reduced from 250 to 200. Essentially, those auto-encoders layers extract appropriate features in an unsupervised manner. The output is fed to Softmax Layer for multi class classification that helps to generate the classification probabilities. The whole architecture, on the other hand, propagates the error by using a backpropagation algorithm. This allows the framework to have supervised characteristics as aforementioned. Auto-encoder parameters for this dataset are shown in Table 7.

Table 7. Auto-Encoder Parameters for Diagnosis of Cardiac Arrhythmia Using Post-Processing Technique.

Parameter	First Auto-Encoder	Second Auto-Encoder
Hidden Layer Size (HS)	250	200
Max Epoch Number (MEN)	130	109
L2 Weight Regularization	0.003	0.001
Sparsity Regularization (SR)	3	1
Sparsity Proportion (SP)	0.12	0.1

Table 8 presents the parameters of PSO which are employed to estimate the best parameters of the linear model. Table 9 demonstrates the proposed framework experimental performance regarding the performance evaluation parameters.

Table 8. PSO Parameters for Diagnosis of Cardiac Arrhythmia.

PSO Parameter	Value
Number of particles	60
Maximum iteration	45
Cognitive parameter	2
Social parameter	2
Min inertia weight	0.9
Max inertia weight	0.2

Table 9. Diagnosis of Cardiac Arrhythmia Results.

Parameter	DSAEs without Post-Processing	DSAEs Using PSO
Recall	0.7843	0.9959
TNR	0.8667	0.9904
Precision	0.8000	0.9918
NPV	0.8553	0.9952
ACC	0.8333	0.9934
F1-s	0.7921	0.9939
MCC	0.6531	0.9866

For this dataset, the linear model parameters are estimated in almost 28 epochs and produce 2.11 error rate as illustrated in Figure 6.

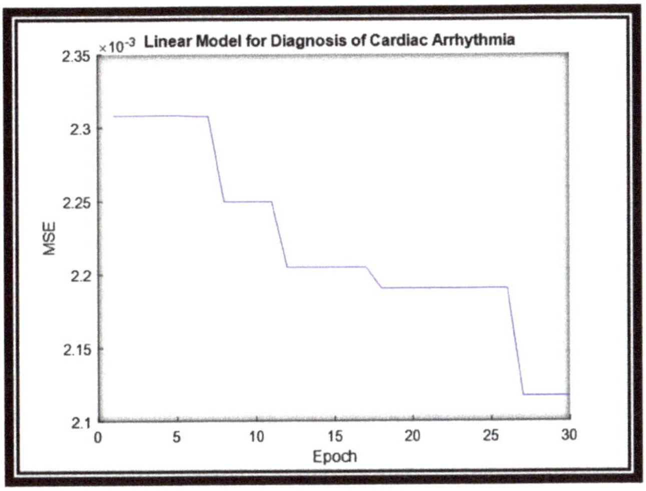

Figure 6. The MSE for the Linear System for Diagnosis of Cardiac Arrhythmia.

4.4. Statistical Significance Analysis of Algorithms in the Proposed Method

In applied machine learning, comparing the algorithms and proposing a final appropriate model for the presented problem is a common approach. Models are generally evaluated using resampling methods (k-fold cross-validation etc.). In these methods, mean performance scores are calculated and compared directly. This approach can give wrong ideas because it is difficult to understand whether the difference between mean performance scores is real or the result of a statistical chance. Statistical significance tests are proposed to overcome this problem and measure the likelihood of the samples with the assumption that they were selected from the equivalent distribution. If this assumption, or null hypothesis, is rejected (if a critical value is smaller than the significance level), it suggests that the difference in skill scores is statistically significant.

Once the data is distributed normally, the two-sample t-test (regarding independent sets) and the paired t-test (for matched samples) are possibly considered the most extensively preferred methods in statistics for the assessment of differences between two samples [53]. A t-test is a type of statistical test that is employed to compare the means of two groups. A 2-tailed paired t-test is preferred in this study to compare the difference between the results without post-processing using PSO and the results after post-processing with PSO (Figures 7–9) in order to evaluate if there is a statistically significant difference when the results are optimized. Two-tailed tests are able to identify differences in either path, greater or less than [54].

A 2-tailed paired *t*-test is applied in Excel on the two matched groups of epileptic seizure detection and *p*-value is calculated as 0.002463, that is, less than the standard level of significance ($p < 0.05$) so a statistically significant difference is noted on this data without using PSO and using PSO. The null hypothesis can be rejected since the sample data support the hypothesis that the population means are dissimilar.

A 2-tailed paired *t*-test is applied in Excel on the two matched groups of SPECTF classification and *p*-value is calculated as 0.020919, that is, less than the standard level of significance ($p < 0.05$) so a statistically significant difference is noted on this data without using PSO and using PSO. The null hypothesis can be rejected since the sample data support the hypothesis that the population means are dissimilar.

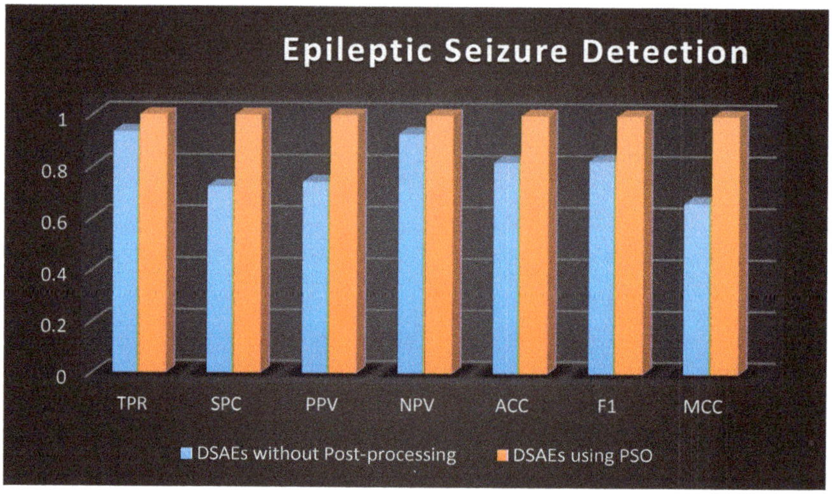

Figure 7. Graphical Representation of Performance Criteria for Epileptic Seizure Detection.

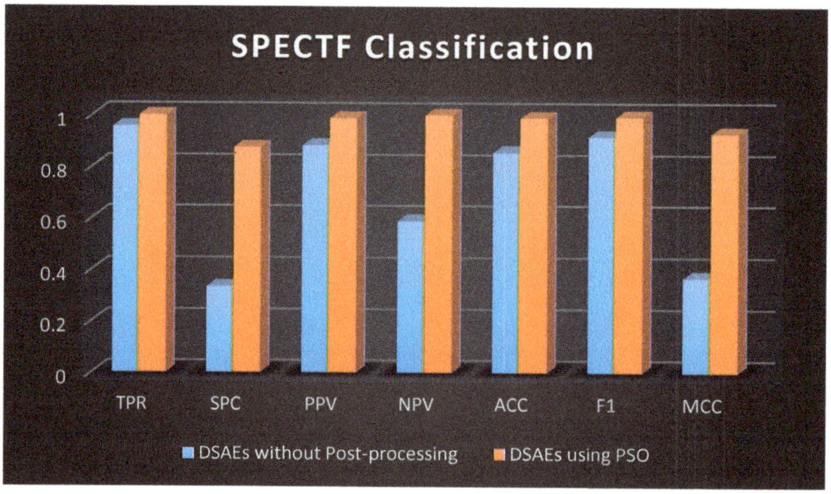

Figure 8. Graphical Representation of Performance Criteria for SPECTF Classification.

A 2-tailed paired *t*-test is applied in Excel on the two matched groups of diagnosis of cardiac arrhythmia and *p*-value is calculated as 0.000307, that is, less than the standard level of significance ($p < 0.05$) so a statistically significant difference is noted on this data without using PSO and using PSO. The null hypothesis can be rejected since the sample data support the hypothesis that the population means are dissimilar.

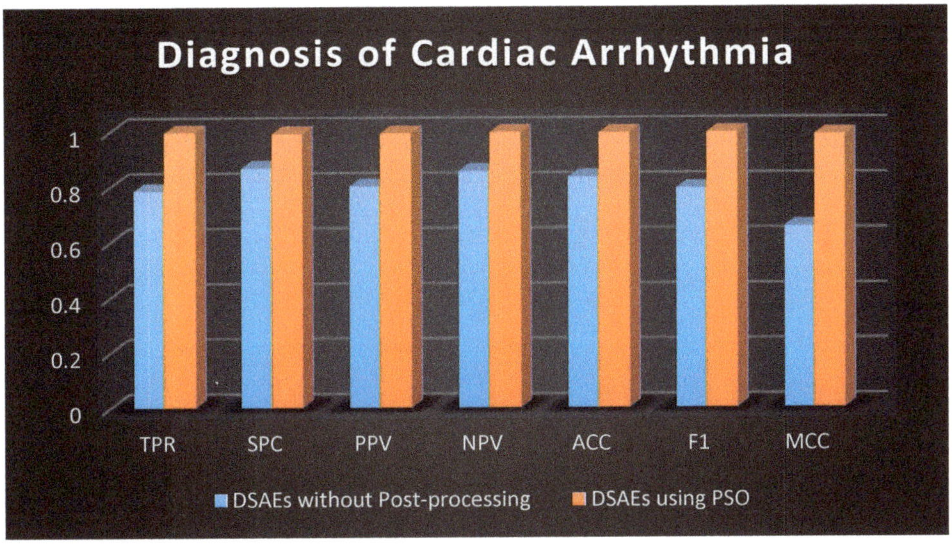

Figure 9. Graphical Representation of Performance Criteria for Diagnosis of Cardiac Arrhythmia.

4.5. Performance Evaluation of the Framework Using Benchmark Datasets

The results of the proposed method, performed on benchmark datasets, are compared to several studies presented in this field. Then, the previous studies are analyzed to reveal the performance of the proposed framework. The comparison results for each dataset are detailed in Tables 10–12. Table 10 represents the comparison between the proposed framework and the leading state-of-the-art studies using Epileptic Seizure Dataset [51], whereas Table 11 involves the comparison based on SPECTF Dataset). Table 12, on the other hand, represents the performance comparison using Cardiac Arrhythmia Dataset. Details of both SPECTF and Cardiac Arrhythmia Datasets can be seen in [52].

Table 10. Evaluation of the Proposed Framework with Leading State-of-the art Studies for Epileptic Seizure Detection.

Reference	Method	Accuracy
[36]	Time–frequency domain feature-RNN	99.6%
[17]	WT + ANN	92.0%
[18]	Discrete WT-mixture of expert model	94.5%
[19]	Entropy measures-ANFIS	92.22%
[20]	Time–frequency analysis—ANN	100%
[21]	Fast Fourier transform-DT	98.72%
[22]	WPD-PCA-GMM	99.00%
[23]	Entropies + HOS + Higuchi FD + Hurst exponent + FC	99.70%
[24]	DTCWT + CVANN-3	100%
[25]	Deep auto-encoder using Taguchi method	100%
[26]	Deep Auto-Encoder + Energy Spectral Density	100%
Proposed Framework	**Deep auto-encoder and linear model based PSO**	**100%**

Table 11. Comparison of SPECTF Classification Results.

Reference	Method	Accuracy
[38]	SVDD	82.7%
[39]	SVDD-based outlier detection	90%
[37]	K2 SDBNS ECFBN	94.03% 95.59% 95.76%
[55]	mc-MKC mc-SVM	79.9% 79.1%
[40]	TCM-IKN N	90%
[41]	C-GAME + Johnson + c4.5 RMEP + Johnson + c4.5	84.4% 81.7%
[16]	Sparsity-based dictionary learning + SVM	97.8%
[26]	Deep Auto-Encoder + Energy Spectral Density	96.79%
Proposed Framework	**Deep auto-encoder and linear model based PSO**	**98.93%**

Table 12. Comparison the performance of the framework on Cardiac Arrhythmia Dataset.

Reference	Method		Accuracy
	Feature Extraction Technique	**Classifier**	
[27]	Enhanced F-score and sequential forward search	k-NN SVM	74% 69%
[28]	Wrapper method	MLP k-NN SVM	78.26% 76.6% 74.4%
[29]	PCA	Kernel difference weighted k-NN	70.66%
[30]	-	MLP+ Static backpropagation algorithm	86.67%
[31]	Best First and CsfSubsetEval	RBF	81%
[32]	-	Modular neural network model	82.22%
[33]	-	ANN models + Static backpropagation algorithm + momentum learning rule	86.67%
[34]	One-against-all	SVM	73.40%
[35]	-	Resampling strategy based random forest (RF) ensemble classifier	90%
[26]	Energy Spectral Density + Deep Auto-Encoders	Softmax	99.1%
Proposed Framework	Deep auto-encoder and linear model based PSO	Softmax	99.27%

4.5.1. Epileptic Seizure Dataset

According to the results shown in Table 10, the proposed framework presented better results than a number of studies [17–19,21–23,36] and presented the same results as other studies with a difference in the complexity and execution time. Peker et al. [24] propose traditional machine techniques which require a long processing time when compared with our proposed framework exactly in high-dimensional features such as epileptic seizure detection. Moreover, in a recent study, the authors propose to train DAEs using the Taguchi method for complex systems. According to this, the parameters are fitted manually when compared with our proposed framework that automatically optimizes the obtained results without needing to repeat experiments manually to obtain the best accuracy [25].

4.5.2. SPECTF Dataset

For this sub-section, results obtained from the proposed framework are compared with well-known studies in the field of SPECTF classification, as shown in Table 11.

The proposed framework achieves better outcomes than all studies can be seen in [16,37–41,55].

4.5.3. Cardiac Arrhythmia Dataset

Finally, the proposed framework shows remarkable results when compared with well-known studies in the field of cardiac arrhythmia, as illustrated in Table 12.

Those studies can be seen in [28–35]. The results verify the advantage of the proposed system over previous relevant papers using the Cardiac Arrhythmia dataset. As previously mentioned, there exist 16 different classes for labelling the dataset. Accordingly, the proposed method accomplishes the best result when it is compared with the state-of-the-art studies.

5. Conclusions

This paper proposes a framework for data classification problems. This novel framework incorporates an efficient deep learning approach (DAE) and linear model trained by a metaheuristic algorithm (PSO). Despite their efficiency, DAEs may produce low performance when employed for complex problems, such as EEG signal classification and motion estimation. Accordingly, the overall goal of this framework is to increase the performance of the DAEs by integrating a post processing layer. This layer essentially optimizes the results obtained from DAEs based on a linear model trained by PSO algorithm. This metaheuristic approach is mainly employed to estimate the parameters of the linear model. As it has produced satisfactory results in various problems, it should be noted that it is easy to implement and involves quite a few parameters for tuning.

Experimental results reveal that the proposed framework presents a number of advantages when compared with previous studies in the literature: learning using less data than other methods. The use of deep learning techniques leads to speeding up the processing time in high-dimensional features because it uses greedy layers as compared to convolutional techniques. The framework also proves that the overall performance of DAEs on complex problems can be enhanced by integrating a post processing layer. According to the results obtained, it is concluded that the introduced framework shows favorable results and can be adapted by researchers for any type of data classification problem. Additionally, as a future work, nonlinear and dynamic linear model systems can be proposed as a post-processing technique for enhancing the classification accuracy of the proposed framework. Moreover, additional optimization algorithms can be employed to train the models instead of PSO, such as the genetic algorithm, the gray-wolf optimization algorithm, the bat algorithm, and other classification models can be combined with linear and nonlinear models, such as support vector machines, naive Bayes or decision trees.

Author Contributions: Conceptualization, A.M.K. and H.K., software development, A.M.K. and M.S.G., formal analysis, F.V.Ç. and M.R.T., supervision, F.V.Ç. and M.R.T., validation, A.M.K., M.S.G. and A.M., original draft preparation, A.M.K. and H.K., writing—review and editing, M.S.G. and A.M. All authors have read and agreed to the published version of the manuscript.

Funding: This research received no external funding.

Acknowledgments: This study is continuation of authors' previous studies entitled with "A New Generalized Deep Learning Framework Combining Sparse Auto-encoder and Taguchi Method for Novel Data Classification and Processing" and "A new framework using deep auto-encoder and energy spectral density for medical waveform data classification and processing".

Conflicts of Interest: The authors declare no conflict of interest.

Appendix A

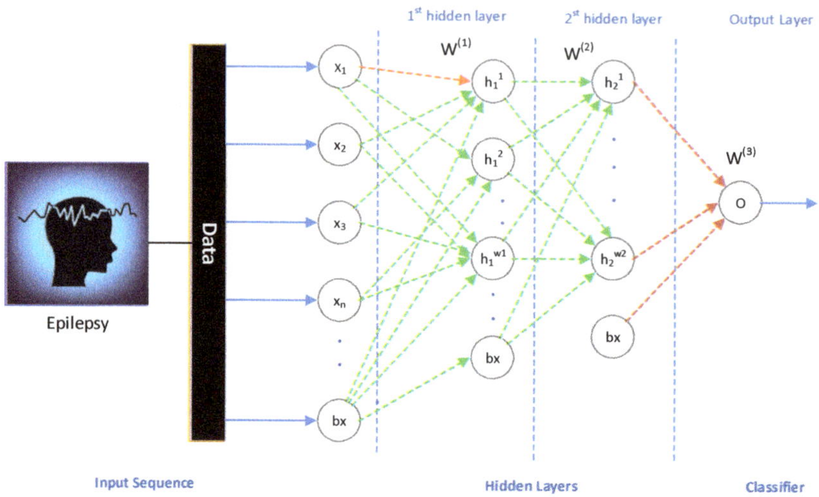

Figure A1. The model of stack a Stacked Sparse Auto-encoder (SSAE) with two hidden layers and a classifier (SoftMax).

References

1. Xu, M.; Fralick, D.; Zheng, J.Z.; Wang, B.; Tu, X.M.; Feng, C. The Differences and Similarities between Two-Sample *t*-test and Paired *t*-test. *Shanghai Arch. Psychiatry* **2017**, *29*, 184–188. [PubMed]
2. Sze, V.; Chen, Y.-H.; Yang, T.-J.; Emer, J.S. Efficient Processing of Deep Neural Networks: A Tutorial and Survey. *Proc. IEEE* **2017**, *105*, 2295–2329. [CrossRef]
3. Luckow, A.; Cook, M.; Ashcraft, N.; Weill, E.; Djerekarov, E.; Vorster, B. Deep learning in the automotive industry: Applications and tools. In Proceedings of the 2016 IEEE International Conference on Big Data (Big Data), Washington, DC, USA, 5–8 December 2016; pp. 3759–3768. [CrossRef]
4. Memisevic, R. Deep learning: Architectures, algorithms, applications. In Proceedings of the 2015 IEEE Hot Chips 27 Symposium (HCS), Cupertino, CA, USA, 22–25 August 2015; pp. 1–127.
5. Chu, L.W. Alzheimer's disease: Early diagnosis and treatment. *Hong Kong Med. J.* **2012**, *18*, 228–237. [PubMed]
6. Pushkar, B.; Paul, M. Early Diagnosis of Alzheimer's Disease: A Multi—Class Deep Learning Framework with Modified k- sparse Autoencoder Classification. In Proceedings of the 2016 International Conference on Image and Vision Computing New Zealand (IVCNZ), Palmerston North, New Zealand, 21–22 November 2016.
7. Tong, H.; Liu, B.; Wang, S. Software defect prediction using stacked denoising autoencoders and two-stage ensemble learning. *Inf. Softw. Technol.* **2018**, *96*, 94–111. [CrossRef]
8. Kuo, J.Y.; Pan, C.W.; Lei, B. Using stacked denoising autoencoder for the student droupout predication. In Proceedings of the 2017 IEEE International Symposium on Multimedia (ISM), Taichung, Taiwan, 11–13 December 2017; pp. 483–488.
9. Xiong, Y.; Zuo, R. Recognition of geochemical anomalies using a deep autoencoder network. *Comput. Geosci.* **2016**, *86*, 75–82. [CrossRef]
10. Salaken, S.M.; Khosravi, A.; Khatami, A.; Nahavandi, S.; Hosen, M.A. Lung cancer classification using deep learned features on low population dataset. In Proceedings of the 2017 IEEE 30th Canadian Conference on Electrical and Computer Engineering (CCECE), Windsor, ON, Canada, 30 April–3 May 2017; pp. 1–5.
11. Khatab, Z.E.; Hajihoseini, A.; Ghorashi, S.A. A Fingerprint Method for Indoor Localization Using Autoencoder Based Deep Extreme Learning Machine. *IEEE Sens. Lett.* **2017**, *2*, 1–4. [CrossRef]

12. Khan, U.M.; Kabir, Z.; Hassan, S.A.; Ahmed, S.H. A Deep Learning Framework Using Passive Wi-Fi Sensing for Respiration Monitoring. In Proceedings of the GLOBECOM 2017—2017 IEEE Global Communications Conference, Singapore, 4–8 December 2017; pp. 1–6.
13. Tang, X.-S.; Hao, K.; Wei, H.; Ding, Y. Using line segments to train multi-stream stacked autoencoders for image classification. *Pattern Recognit. Lett.* **2017**, *94*, 55–61. [CrossRef]
14. Yin, C.; Zhu, Y.; Fei, J.; He, X. A Deep Learning Approach for Intrusion Detection Using Recurrent Neural Networks. *IEEE Access* **2017**, *5*, 21954–21961. [CrossRef]
15. Yu, Z.; Tan, E.-L.; Ni, D.; Qin, J.; Chen, S.; Li, S.; Lei, B.; Wang, T. A Deep Convolutional Neural Network-Based Framework for Automatic Fetal Facial Standard Plane Recognition. *IEEE J. Biomed. Health Inform.* **2017**, *22*, 874–885. [CrossRef]
16. Srinivas, M.; Bharath, R.; Rajalakshmi, P.; Mohan, C.K. Multi-level classification: A generic classification method for medical datasets. In Proceedings of the 2015 17th International Conference on E-health Networking, Application & Services (HealthCom), Boston, MA, USA, 14–17 October 2015; pp. 262–267. [CrossRef]
17. Subasi, A.; Erçelebi, E. Classification of EEG signals using neural network and logistic regression. *Comput. Methods Programs Biomed.* **2005**, *78*, 87–99. [CrossRef] [PubMed]
18. Subasi, A. EEG signal classification using wavelet feature extraction and a mixture of expert model. *Expert Syst. Appl.* **2007**, *32*, 1084–1093. [CrossRef]
19. Kannathal, N.; Choo, M.L.; Acharya, U.R.; Sadasivan, P. Entropies for detection of epilepsy in EEG. *Comput. Methods Programs Biomed.* **2005**, *80*, 187–194. [CrossRef]
20. Tzallas, A.T.; Tsipouras, M.G.; Fotiadis, D.I. Automatic Seizure Detection Based on Time-Frequency Analysis and Artificial Neural Networks. *Comput. Intell. Neurosci.* **2007**, *2007*, 80510. [CrossRef] [PubMed]
21. Polat, K.; Güneş, S. Classification of epileptiform EEG using a hybrid system based on decision tree classifier and fast Fourier transform. *Appl. Math. Comput.* **2007**, *187*, 1017–1026. [CrossRef]
22. Acharya, U.R.; Sree, S.V.; Alvin, A.P.C.; Suri, J.S. Use of principal component analysis for automatic classification of epileptic EEG activities in wavelet framework. *Expert Syst. Appl.* **2012**, *39*, 9072–9078. [CrossRef]
23. Acharya, U.R.; Sree, S.V.; Ang, P.C.A.; Yanti, R.; Suri, J.S. Application of non-linear and wavelet based features for the automated identification of epileptic EEG signals. *Int. J. Neural Syst.* **2012**, *22*, 1250002. [CrossRef] [PubMed]
24. Peker, M.; Şen, B.; Delen, D. A Novel Method for Automated Diagnosis of Epilepsy Using Complex-Valued Classifiers. *IEEE J. Biomed. Heal. Inform.* **2015**, *20*, 108–118. [CrossRef]
25. Karim, A.M.; Güzel, M.S.; Tolun, M.R.; Kaya, H.; Çelebi, F.V. A New Generalized Deep Learning Framework Combining Sparse Autoencoder and Taguchi Method for Novel Data Classification and Processing. *Math. Probl. Eng.* **2018**, *2018*, 3145947. [CrossRef]
26. Karim, A.M.; Serdar, G.M.; Tolun, M.R.; Kaya, H.; Çelebi, F.V. A new framework using deep auto-encoder and energy spectral density for medical waveform data classification and processing. *Biocybern. Biomed. Eng.* **2019**, *39*, 148–159. [CrossRef]
27. Niazi, K.A.K.; Khan, S.A.; Shaukat, A.; Akhtar, M. Identifying best feature subset for cardiac arrhythmia classification. In Proceedings of the 2015 Science and Information Conference (SAI), London, UK, 28–30 July 2015; pp. 494–499.
28. Mustaqeem, A.; Anwar, S.M.; Majid, M.; Khan, A.R. Wrapper method for feature selection to classify cardiac arrhythmia. In Proceedings of the 39th Annual International Conference of the IEEE Engineering in Medicine and Biology Society (EMBC), Seogwipo, Korea, 11–15 July 2017; pp. 3656–3659. [CrossRef]
29. Zuo, W.; Lu, W.; Wang, K.; Zhang, H. Diagnosis of cardiac arrhythmia using kernel difference weighted KNN classifier. *Comput. Cardiol.* **2008**, *35*, 253–256. [CrossRef]
30. Jadhav, S.M.; Nalbalwar, S.L.; Ghatol, A.A. ECG arrhythmia classification using modular neural network model. In Proceedings of the IEEE EMBS Conference on Biomedical Engineering and Sciences (IECBES), Kuala Lumpur, Malaysi, 30 November–2 December 2010; pp. 62–66.
31. Persada, A.G.; Setiawan, N.A.; Nugroho, H. Comparative study of attribute reduction on arrhythmia classification dataset. In Proceedings of the International Conference on Information Technology and Electrical Engineering (ICITEE), Yogyakarta, Indonesia, 7–8 October 2013; pp. 68–72. [CrossRef]

32. Jadhav, S.M.; Nalbalwar, S.L.; Ghatol, A. Artificial Neural Network based cardiac arrhythmia classification using ECG signal data. In Proceedings of the International Conference on Electronics and Information Engineering, Kyoto, Japan, 1–3 August 2010; Volume 1, pp. V1–V228. [CrossRef]
33. Jadhav, S.M.; Nalbalwar, S.L.; Ghatol, A.A. Artificial Neural Network Based Cardiac Arrhythmia Disease Diagnosis. In Proceedings of the International Conference on Process. Automation, Control. and Computing, Coimbatore, India, 20–22 July 2011; pp. 1–6. [CrossRef]
34. Kohli, N.; Verma, N.K.; Roy, A. SVM based methods for arrhythmia classification in ECG. In Proceedings of the International Conference on Computer and Communication Technology (ICCCT), Allahabad, India, 17–19 September 2010; pp. 486–490. [CrossRef]
35. Özçift, A. Random forests ensemble classifier trained with data resampling strategy to improve cardiac arrhythmia diagnosis. *Comput. Biol. Med.* **2011**, *41*, 265–271. [CrossRef]
36. Srinivasan, V.; Eswaran, C.; Sriraam, A.N. Artificial Neural Network Based Epileptic Detection Using Time-Domain and Frequency-Domain Features. *J. Med. Syst.* **2005**, *29*, 647–660. [CrossRef]
37. Wei, J.; Yu, H.; Wang, J. The research of Bayesian method from small sample of high-dimensional dataset in poison identification. In Proceedings of the IEEE 4th International Conference on Software Engineering and Service Science, Beijing, China, 23–25 May 2013; pp. 705–709.
38. Cha, M.; Kim, J.S.; Baek, J.-G. Density weighted support vector data description. *Expert Syst. Appl.* **2014**, *41*, 3343–3350. [CrossRef]
39. Liu, B.; Xiao, Y.; Cao, L.; Hao, Z.; Deng, F. SVDD-based outlier detection on uncertain data. *Knowl. Inf. Syst.* **2013**, *34*, 597–618. [CrossRef]
40. Cui, L.-l.; Zhu, H.-c.; Zhang, L.-k.; Luan, R.-p. Improved kNearest Neighbors Transductive Confidence Machine for Pattern Recognition. *IEEE Int. Conf. Comput. Des. Appl.* **2010**, *3*, 172–176.
41. Tian, D.; Zeng, X.-J.; Keane, J. Core-generating approximate minimum entropy discretization for rough set feature selection in pattern classification. *Int. J. Approx. Reason.* **2011**, *52*, 863–880. [CrossRef]
42. Zeng, N.; Zhang, H.; Song, B.; Liu, W.; Li, Y.; Dobaie, A.M. Facial expression recognition via learning deep sparse autoencoders. *Neurocomputing* **2018**, *273*, 643–649. [CrossRef]
43. Kennedy, J.; Eberhart, R. Particle swarm optimization. In Proceedings of the IEEE International Conference on Neural Networks (ICNN'95), Perth, Australia, 27 November–1 December 1995; Volume 4, pp. 1942–1948.
44. Harman, R. *A Very Brief Introduction to Particle Swarm Optimization*; Technical Report; Department of Applied Mathematics and Statistics: Bratislava, Slovakia, 1995; pp. 1–4.
45. Kaveh, A.; Nasrollahi, A. A new probabilistic particle swarm optimization algorithm for size optimization of spatial truss structures. *Int. J. Civ. Eng.* **2014**, *12*, 1–13.
46. Ding, W.; Lin, C.-T.; Cao, Z. Deep Neuro-Cognitive Co-Evolution for Fuzzy Attribute Reduction by Quantum Leaping PSO With Nearest-Neighbor Memeplexes. *IEEE Trans. Cybern.* **2018**, *49*, 2744–2757. [CrossRef] [PubMed]
47. Serdar, G.M.; Kara, M.; Beyazkılıç, M.S. An adaptive framework for mobile robot navigation. *Adapt. Behav.* **2017**, *25*, 30–39. [CrossRef]
48. Rizvi, S.Z.; Abbasi, F.; Velni, J.M. Model Reduction in Linear Parameter-Varying Models using Autoencoder Neural Networks. In Proceedings of the Annual American Control Conference (ACC), Milwaukee, WI, USA, 27 June 2018; pp. 6415–6420.
49. Siswantoro, J.; Prabuwono, A.S.; Abdullah, A.; Idrus, B. A linear model based on Kalman filter for improving neural network classification performance. *Expert Syst. Appl.* **2016**, *49*, 112–122. [CrossRef]
50. Noy, D.; Menezes, R. Parameter estimation of the Linear Phase Correction model by hierarchical linear models. *J. Math. Psychol.* **2018**, *84*, 1–12. [CrossRef]
51. Andrzejak, R.G.; Lehnertz, K.; Mormann, F.; Rieke, C.; David, P.; Elger, C.E. Indications of nonlinear deterministic and finite-dimensional structures in time series of brain electrical activity: Dependence on recording region and brain state. *Phys. Rev. E* **2001**, *64*, 061907. [CrossRef]
52. Dua, D.; Karra, T. *Machine Learning Repository*; School of Information and Computer Sciences, University of California: Irvine, CA, USA, 2017; Available online: http://archive.ics.uci.edu/ml (accessed on 12 July 2019).
53. Xu, G.; Fang, W. Shape retrieval using deep autoencoder learning representation. In Proceedings of the 13th International Computer Conference on Wavelet Active Media Technology and Information Processing (ICCWAMTIP), Chengdu, China, 16–18 December 2016; pp. 227–230. [CrossRef]

54. Kim, T.K. T test as a parametric statistic. *Korean J. Anesthesiol.* **2015**, *68*, 540–546. [CrossRef] [PubMed]
55. Kumar, R.; Chen, T.; Hardt, M.; Beymer, D.; Brannon, K.; Syeda-Mahmood, T. Multiple Kernel Completion and its application to cardiac disease discrimination. In Proceedings of the IEEE 10th International Symposium on Biomedical Imaging, San Francisco, CA, USA, 7 April 2013; pp. 764–767.

Publisher's Note: MDPI stays neutral with regard to jurisdictional claims in published maps and institutional affiliations.

sensors

Article

Contribution to Speeding-Up the Solving of Nonlinear Ordinary Differential Equations on Parallel/Multi-Core Platforms for Sensing Systems

Vahid Tavakkoli *, Kabeh Mohsenzadegan, Jean Chamberlain Chedjou and Kyandoghere Kyamakya

Institute for Smart Systems Technologies, University Klagenfurt, A9020 Klagenfurt, Austria; kabehmo@edu.aau.at (K.M.); Jean.Chedjou@aau.at (J.C.C.); kyandoghere.kyamakya@aau.at (K.K.)

* Correspondence: vtavakko@edu.aau.at; Tel.: +43-463-2700-3540

Received: 18 September 2020; Accepted: 26 October 2020; Published: 28 October 2020

Abstract: Solving ordinary differential equations (ODE) on heterogenous or multi-core/parallel embedded systems does significantly increase the operational capacity of many sensing systems in view of processing tasks such as self-calibration, model-based measurement and self-diagnostics. The main challenge is usually related to the complexity of the processing task at hand which costs/requires too much processing power, which may not be available, to ensure a real-time processing. Therefore, a distributed solving involving multiple cores or nodes is a good/precious option. Also, speeding-up the processing does also result in significant energy consumption or sensor nodes involved. There exist several methods for solving differential equations on single processors. But most of them are not suitable for an implementation on parallel (i.e., multi-core) systems due to the increasing communication related network delays between computing nodes, which become a main and serious bottleneck to solve such problems in a parallel computing context. Most of the problems faced relate to the very nature of differential equations. Normally, one should first complete calculations of a previous step in order to use it in the next/following step. Hereby, it appears also that increasing performance (e.g., through increasing step sizes) may possibly result in decreasing the accuracy of calculations on parallel/multi-core systems like GPUs. In this paper, we do create a new adaptive algorithm based on the Adams–Moulton and Parareal method (we call it PAMCL) and we do compare this novel method with other most relevant implementations/schemes such as the so-called DOPRI5, PAM, etc. Our algorithm (PAMCL) is showing very good performance (i.e., speed-up) while compared to related competing algorithms, while thereby ensuring a reasonable accuracy. For a better usage of computing units/resources, the OpenCL platform is selected and ODE solver algorithms are optimized to work on both GPUs and CPUs. This platform does ensure/enable a high flexibility in the use of heterogeneous computing resources and does result in a very efficient utilization of available resources when compared to other comparable/competing algorithm/schemes implementations.

Keywords: ODE Solver; OpenCL; Parareal; parallel/multi-core computing; sensing systems; heterogenous embedded systems

1. Introduction

The history of using differential equations has traces in calculus from the old Newton's times. Since then it has evolved so much, and it is extensively used in many different branches of science and engineering. The numerical solving of differential equations with initial conditions is a classic problem, which has emerged before the computer invention and has various different usages in physics [1], engineering [2], chemistry [3], economics [4], biology [5], and several other disciplines.

Specifically in sensors, ODEs are involved in various processing endeavors such as to detect anomalies in machines related sensor data [6], or to model nonlinear sensors like time-variant inductors [7] or piezoelectrically actuated microcantilever sensors [8], and/or to study sensors' behavior or to optimize sensors' performance. ODEs also used to find sensor's optimal location [9]. Quorum sensing (QS), which is based on bacterial communication, can also be modeled with differential equations. Furthers, ODEs can also be used in self-organizing networks, self-diagnostic and environmental monitoring systems [10]. Hence, finding new ways for solving ODEs in shorter time can help to save, besides processing related energy consumption, both money and time, while using cheaper devices for better performing applications.

Differential equations have many different forms. In this paper, we do focus on ordinary differential equations (ODEs).

Equation (1) is showing an example of this type of differential equations:

$$\dot{y}(t) = f(t, y(t));\ y(t_0) = y_0,\ t \in [t_0, t_e] \tag{1}$$

where y is a vector valued function of t (time), $y(t) : \mathbb{R} \rightarrow \mathbb{R}^n$, n is dimension of problem, the time derivative of y, $\dot{y}(t)$, is a function of y and t, and the function f has the values domain $f : \mathbb{R}^n \rightarrow \mathbb{R}^n$. Further, $y(t_0) = y_0$ is called the initial value. Thus, we do have a so-called initial value problem (IVP)) and y_0 is the starting point for calculations at t_0. The solving of Equation (1) shall calculate the values of y from t_0 until t_e.

We need to determine the problem's solutions (i.e., y(t)) for all values of t within the interval $[t_0, t_e]$, this thus starting from the initial value $y(t_0)$ up to the final value of the $y(t_e)$. The solution of Equation (1) can be found by applying various appropriate methods, which are either numerical or analytical. For those cases for which it is hard to find an analytical solution, one does usually then involve numerical methods.

Regarding numerical methods, the simplest way to solve Equation (1) is to integrate the function f for over the study area (i.e., $[t_0, t_e]$, provided the function f does satisfy the so-called Lipchitz conditions. A numerical solving can be implemented through a discretized version of Equation (1), which is given in Equation (2). In Equation (2), y_{t+1} is the result of the calculation of one step. The calculation of one step is obtained by taking the previous value y_t plus the integral of $f(t, y_t)$ from the previous time t up to the current time $t + 1$. For calculating the integral, various traditional methods like Euler, Runge-Kutta, etc. can be used:

$$y_{t+1} = y_t + \int_t^{t+1} f(s, y_s) \cdot ds \tag{2}$$

where s is time between t and $t + 1$, and y_s is value of function f in time s.

In the case of a single computing core, there is no problem to achieve an efficient usage of computing resources. In this one-core context, it is very easy and straight-forward to implement Equation (2) and, after the calculation of one step is finished, one does move on calculating the next step. The steps are solved in a serial manner and the result of each step is then be used for next steps' calculations. However, when one works on a multi-core/parallel platform, that sequential model cannot be used anymore, as other available resources/cores/nodes would have nothing to do.

There exists five different space-time parallel computing methods/schemes for implementing the difference equation Equation (2) [10]. Those five schemes are the following ones: domain decomposing, parallel solver, multiple shooting, direct time parallel, and multi grid.

In the domain decomposing scheme, one does separate, if possible, the problem into n sub-problems and solve each of them separately. This can be realized by integrating the 'domain decomposition' and the so-called waveform relaxation [11,12]. Basically, in this solver type, the problem domain is decomposed into overlapping sub-domains, and each domain is then solved separately [13]. Choosing the correct way for decomposing is very important for increasing the overall performance. Also, the 'decomposing method' can be varied due to the nature of Equation (1) [14,15].

A further approach is to use a parallel integration method. This is however not possible for single-step integration methods. See for example Equation (3), where the Euler method is presented. As one can see, for calculating y_t, one step is required, and this step cannot be separated (broken down) into smaller tasks/sub-steps for a separate implementation on different cores of a parallel system. The Δt is the calculation step. A lower value of Δt provides a higher accuracy for solving a given problem but it does however thereby increase the resulting calculation time.

$$y_t = \Delta t \cdot \dot{y}(t) + y_{t-1} \tag{3}$$

Therefore, this above-named further approach, i.e., a "parallel integration method", is only possible while using multi-steps methods such as Runge-Kutta or Adams-Bashforth integrators, which are classified as larger group of Generalized Linear Model (GLM) solvers. GLM solvers are explained in detail in Section 2.1.

Methods like Runge-Kutta are multi-steps iterated methods [16–19]; this means one can distribute calculations of each step on different computing nodes. But at the end of each step, the different computing nodes should send their results to one node to sum-up or combine them appropriately and then calculate a new value. This last part of the lastly described scheme does visibly create a bottleneck w.r.t. to the potential speeding-up of the solving of differential equations while using a multi-step method.

In the shooting methods which were introduced by Nievergelt in 1964 [20], Equation (2) is decomposed in the time direction into semi-linear boundary value problems. Smaller problems are then solved with higher accuracy in parallel, but the error will then be corrected in a serial operation. Although, this method is by definition sequential because of the integrated serial error correction. However, it normally does cost much less than the high-accuracy calculation of results on hole of integration area. Therefore, this brings real advantages in the perspective of solving any ODE problem [21].

Multigrid methods like the so-called "domain decomposition" can be used for solving non-linear ODE's. The problem is discretized with finite approximations into sparse linear systems of equations. This linear system is later solved via stationary iterative schemes such as the Gauss-Seidel method [22,23].

In lastly described approach, one tries to solve the problem directly without any iteration. All iterations for solving n points will be put in one place in one matrix and the problem is then solved together at once [24].

Furthers, it can be observed that several scientific works have been undertaken in order to create new integration methods, which can provide ODE solvers with better possibilities for an efficient implementation on parallel platforms. These efforts mostly focused on creating the so-called Adams-Bashforth derivative methods such as parallel Adam-Bashford (PAB) and parallel Adam-Moulton (PAM) [25,26]. These last methods have shown very good scalability performance while increasing the number of computing nodes.

Today, most of modern computers have both n core CPUs (n-CPUs) and GPUs. The increasing power of CPUs and GPUs is mostly reached by increasing the number of computing nodes. Although the number of computing nodes has significantly increased in GPUs but also in n-CPUs, the need for algorithms capable to efficiently use the multi-core computing resources is strong. It has been shown that implementing "problem solvers" on parallel/multi-node platforms can speed-up the solving in many scientific fields such as fluid dynamics [27], finite elements methods [28], molecular dynamics research [29], applied physics [30], chemical kinetics [31], etc.

On the other hand, for writing programs which can efficiently run on different computing architectures is not a trivial problem. For solving this concern, some middleware concepts/platforms which do support different types of n-CPUs or GPUs architectures have been developed and introduced. In this paper, we use the so-called OpenCL platform. It is possible, by using OpenCL, to run programs directly on CPU or GPU. However, this programming framework, like other similar frameworks has also its own restrictions. In this paper, we do introduce a new solver type/concept which does well fit and is integrated in the OpenCL platform. This novel ODE solver concept implemented through OpenCL has been extensively tested and benchmarked with other related competing famous/well-known algorithms from the most relevant literature.

This paper does present a very brief critical overview about related works in Section 2. Then, our novel ODE parallel-solver concept is introduced in Section 3. The implementation system architecture in OpenCL, which does support the running application of our novel solver concept is explained in Section 4. Then, extensive experiments and a comprehensive benchmarking are presented and discussed in Sections 5 and 6. To finish, comprehensive concluding remarks, which summarize the quintessence of the results obtained, are presented in Section 7.

2. Related Works

As briefly explained above, we should search for multi-stage methods, which do have the potential for solving each stage of the problem, possibly independently of each other [32]. For this study, the knowingly best-performing multi-step algorithms have been selected for analysis and possibly benchmarking too. One of those methods is derived from the Runge-Kutta family and we call it "Iterated Runge-Kutta". And two further methods are derived the from the Adams–Bashforth family, which are respectively called "Parallel Adams–Bashforth (PAB)" and "Parallel Adams-Moulton (PAM)" [25].

2.1. General Linear Methods

The General Linear Method (GLM) as proposed by Butcher in 1966 was defined to generalize and integrate both Runge-Kutta (multi-stage) methods and linear multistep (multi-value) methods. During each step of the calculation, one considers r numbers of previous values and s stages. At the start of each step, we have input items from the previous steps as follows:

$$y_i^{[n]},\ i = 1,\ 2,\ \ldots,\ r \tag{4}$$

And during calculation of stages in one step, we have stage derivatives as follows:

$$Y_i,\ F_i\ ,\ i = 1,\ 2,\ \ldots,\ s \tag{5}$$

Thus, this method has the following variables for calculating the next stage $n+1$:

$$y_{[n]} = \begin{bmatrix} y_1^{[n]} \\ y_2^{[n]} \\ \vdots \\ y_r^{[n]} \end{bmatrix},\ y_{[n+1]} = \begin{bmatrix} y_1^{[n+1]} \\ y_2^{[n+1]} \\ \vdots \\ y_r^{[n+1]} \end{bmatrix},\ Y = \begin{bmatrix} Y_1 \\ Y_2 \\ \vdots \\ Y_s \end{bmatrix},\ F = \begin{bmatrix} F_1 \\ F_2 \\ \vdots \\ F_s \end{bmatrix} \tag{6}$$

These quantities are related to each other by the following equation, see Equation (7):

$$\begin{aligned} &Y = h(A \otimes I)\, F + (U \otimes I)\, y_{[n]} \\ &y_{[n+1]} = h(B \otimes I)\, F + (V \otimes I) y_{[n]} \\ &F = f(Y) \end{aligned} \tag{7}$$

where $\otimes$ is tensor product, h is the step-size in $[t_n,\ t_{n+1}]$, and A, U, B and V are constant matrices having the following respective dimensions:

$$A \in \mathbb{R}^{s\times s},\ U \in \mathbb{R}^{s\times r},\ B \in \mathbb{R}^{r\times s},\ V \in \mathbb{R}^{r\times r} \tag{8}$$

In Equation (7), the result of the step ($y_{[n+1]}$) is calculated based on the previous values ($y_{[n]}$) and the stage values (F, Y). F is calculated directly from Y based on the Equation (1) definition. For those linear multistep methods for which previous values ($y_{[n]}$) are required, the starting vector can be calculated by one of the nth-order Runge-Kutta methods such as the so-called Dormand-Prince method (DOPRI) which do not need previous values.

As Butcher explained [33], the customization of GLM is creating a different ODE solver, which can be customized to have properties of either the Runge-Kutta method by setting $r = 1$ or the linear multistep method by setting $s = 1$. For solving non-stiff ODE problems, some methods based on GLM have been created by customizing the parameters r and s and/or the matrices A, U, B and V.

For example, the classical fourth order of Runge-Kutta can be expressed in GLM with the following matrices:

$$A = \begin{bmatrix} 0 & 0 & 0 & 0 \\ \frac{1}{2} & 0 & 0 & 0 \\ 0 & \frac{1}{2} & 0 & 0 \\ 0 & 0 & 1 & 0 \end{bmatrix}, \; U = \begin{bmatrix} 1 \\ 1 \\ 1 \\ 1 \end{bmatrix}, \; B = \begin{bmatrix} \frac{1}{6} & \frac{1}{3} & \frac{1}{3} & \frac{1}{6} \end{bmatrix}, \; V = [1] \tag{9}$$

And in the case of the second order Adam-Bashforth method, A, U, B and V can expressed as follows:

$$A = [0], \; U = \begin{bmatrix} 1 & \frac{3}{2} & -\frac{1}{2} \end{bmatrix}, \; B = \begin{bmatrix} 0 \\ 1 \\ 0 \end{bmatrix}, \; V = \begin{bmatrix} 1 & \frac{3}{2} & -\frac{1}{2} \\ 0 & 0 & 0 \\ 0 & 1 & 0 \end{bmatrix} \tag{10}$$

By choosing a strictly diagonal or triangular matrix A, the stage calculation will be decoupled into s independent sub-systems. Therefore, in this case, the implementation of the solver on a parallel/multi-core system is much easier as the stage dependency is thereby significantly reduced.

For example, if we have following A matrix:

$$A = \begin{bmatrix} 0 & & & & & & \\ x & 0 & & & & & \\ x & 0 & 0 & & & & \\ x & 0 & 0 & 0 & & & \\ x & x & x & 0 & 0 & & \\ x & x & x & 0 & 0 & 0 & \\ x & x & x & 0 & 0 & 0 & 0 \end{bmatrix} \tag{11}$$

(2,3,4) Stages and (5,6,7) Stages can be computed concurrently as those stage does not use value from each other, Therefore they can be solved in parallel way. This pattern can be seen in the parallel iterated Runge-Kutta (PIRK) or better in the step-independent methods like the PAB or the PAM [25] methods, where $A = [0]$. In this paper, we do also use the GLM solver to create our new solver by customizing the matrices A, U, B and V.

2.1.1. Parallel Iterated Runge-Kutta

This method is defined according to [34–36] and is also based on the GLM method (Equation (7)). The matrices A and V have the following definition:

$$A = \begin{bmatrix} 1 & 0 & \cdots & 0 \\ \vdots & \vdots & \cdots & \vdots \\ 1 & 0 & \cdots & 0 \end{bmatrix}, \; \mathrm{V} = [1] \tag{12}$$

The U matrix is calculated based on related Runge-Kutta method as explained in the previous sections. This method is a very precise method. But it is not using all resources when we have only one ODE equation. The number of steps can be changed during each iteration. Therefore, one can reach a significant speed-up while solving large problems needing too steps to calculate.

An implementation example of this model can be shown in Figure 1:

For i = 1 to n
    Calculate Y in parallel
    Calculate $y_{[i+1]}$ in sequential

Figure 1. Implementation of the parallel Runge-Kutta algorithm. Stage values (Y) can be calculated in parallel but the step result needs to be calculated sequentially.

Figure 1 is showing the calculation flow of the different steps. The stage values (Y) can be done in a parallel way, but each processing unit needs to exchange information during the processing and at the end of each stage. Again, each node requires to exchange information with another specific node in order to sum up all steps and create the step value ($y_{[i+1]}$).

2.1.2. Parallel Adams-Bashforth

This method was introduced by v.d. Houwen and Messina in 1998 [25]. Since then it has been further developed and optimized to be used in parallel platforms [37]. The Parallel Adams-Bashforth (PAB) is based on the Adams–Bashforth corrector by customizing the GLM with A, U, B and V matrices having the following values:

$$A = [0],\ V = a{\cdot}b^T,\ a = \begin{bmatrix} 1 & 1 & \dots & 1 \end{bmatrix},\ b = \begin{bmatrix} 0 & \dots & 0 & 1 \end{bmatrix} \tag{13}$$

The U matrix is calculated based on the related Adams-Baschforth method explained above in the GLM section. It has been proved that by choosing those matrices in Equation (13), the PAB solver becomes super-convergent to the real solution of an ODE problem. Implementing this method on parallel system is not straight forward and requires a special scheduling. Figure 2 is showing a basic scheduling for running this method on 3 processing units. In each iteration, after find the results ($y_{[i]}$), the F values which are to use for the next iteration will be calculated. Thus, each iteration calculation can be done in a parallel way. But after finishing an iteration, each computing unit should exchange its information with other processors in its respective group of processors. This process will be continued until end of the calculation time (Figure 2).

The PAB method can result in an improvement of the speed-up when compared to the Runge-Kutta method because, here, the communication between nodes can be done only at the beginning and at the end of running a stage. Therefore, it is very efficient to implement the PAB method on a parallel system. On the other hand, if we want to implement this method on an OpenCL platform, we do need a very good synchronization. This because the last node having the larger amount of calculations, the other nodes need to wait until it will finalize its calculations and only then let the other nodes synchronize themselves with latest values.

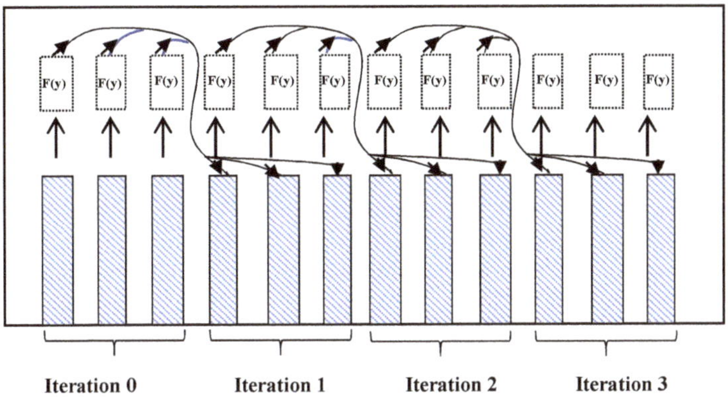

Figure 2. PAB execution and scheduling scheme on 3 processing units. The result value of each iteration is calculated and then the F value for the next iteration is computed. Those values will be propagated to the other processing units for the next iteration. In each iteration, 3 points of the problem are solved.

2.2. Multiple Shotting Methods

In this type of methods, as explained previously in the introduction section, the space-time domain is decomposed into smaller parts (sub-domains) and each subdomain is solved separately. The idea of creating this method is coming from Nievergelt in 1964 [20]. Since then, the method has been developed and extended by different researchers and it is mostly well-known as 'Parareal' algorithm [38–40].

The general implementation of this so-called "Parareal" algorithm is composed/constituted of two propagation operators:

1. The "Coarse approximation," which is $G(t_i, t_{i+1}, y_i)$ with the initial conditions $y_i = y(t_i)$ with the step size h_g.
2. The "Fine approximation," which is $F(t_i, t_{i+1}, y_i)$ with the initial conditions $y_i = y(t_i)$ with the step size h_f.

Where $h_g \gg h_f$, therefore the main difference between the above listed two propagation operators is their respective accuracy and the amount of time they do need to find the result as the coarse approximator has a larger step size.

The main algorithm consists of the following steps [12]:

1. Find the values of $y_1, \ldots, y_n$ by using $y_{i+1} = G(t_i, t_{i+1}, y_i)$ in a sequential way.
2. Copy the $y_1, \ldots, y_n$ values into $g_1, \ldots, g_n$ in parallel.
3. Find the $f_1, \ldots, f_n$ values by using $f_{i+1} = F(t_i, t_{i+1}, y_i)$ in parallel.
4. Update $y_1, \ldots, y_n$ in sequential with the following steps:
   a. $gn_{i+1} = G(t_i, t_{i+1}, y_i)$.
   b. $y_{i+1} = f_{i+1} + gn_{i+1} - g_i$.
5. Copy the $gn_1, \ldots, gn_n$ values into $g_1, \ldots, g_n$.
6. Go to the step 3 until you reach required precision.

This above-described algorithm is also sequential; for each iteration we do also need the values from the previous iteration. Thus, there is no real parallel-time integration as the sequential nature of the process is not removed. But the most expensive part is done in parallel (see Step 3) and solving that most expensive part in parallel will bring a significant advantage when increasing the number of computing nodes. One main disadvantage is, however, that this algorithm needs too many computing nodes to reach a good speed-up. Consequently, it is not efficient to implement it for a small number of computing nodes in the ranges like less than 8 or 16.

2.3. Summary of the Main Previous/Traditional Methods

By comparing the properties of the above presented methods, there is one big difference amongst them. The GLM methods are optimized to be efficiently used in the context of parallel systems' contexts having specified properties like "running schedule" and "number of processing units". We implemented all these three above listed methods on an OpenCL framework in order to carefully analyze their respective performance along with related respective observed implementation restrictions. One restriction (weak point) is related to the poor scalability w.r.t. to the increasing number of cores. One does observe a very poor performance scaling of the algorithms while increasing the number of involved nodes for solving a problem. On the other hand, the Parareal method is showing a very good advantage when increasing number of cores; however, it is showing a very poor scaling performance in presence of a small number of cores (e.g. 8 cores or 16 cores).

This observed gap has motivated us to create a new method based on both the Parallel Adam Bashforth method and the Parareal algorithm by using the advantages of both methods with a significantly higher compatibility with the OpenCL framework.

3. Our Novel Concept, the Parallel Adam-Moulton OpenCL

Our novel method, that we call the Parallel Adam-Moulton OpenCL method (we abbreviate it in PAMCL) is a modified version of the Adams-Bashforth method, which has a scheduling scheme like the PAB method (see Figure 2). This method is defined through the following equation for each group of computing units:

$$A = [0],\ V = a{\cdot}b^T,\ a = \begin{bmatrix} 1 & 2 & \dots & g \end{bmatrix},\ b = \begin{bmatrix} 0 & \dots & 0 & 1 \end{bmatrix} \tag{14}$$

where g is the group size of processors, and the number of previous values is g. Therefore, based on the number of processors in a group, the requirements to the previous values are different. The matrix U is calculated based on the related Adams-Baschforth method as explained in the GLM section.

The starting vector for this algorithm is calculated by a suitable scheme like the DOPRI5 method [41], and each value of F is approximated by using the previous steps of the solution vectors by using the Equation (7). For each calculation stage, we have a $(h \times g)$ step size advantage w.r.t. to h. On the other hand, with a growing size of the "processing units group", we need a higher order method for calculating the result values ($y_{[n+1]}$) by using an Adams-Bashforth algorithm. In this way, we do increase our accuracy without losing in performance.

After calculating the result, we do need a corrector function based on the Adam-Moulton formula to correct the calculations of the previous steps.

Based on both the predictor (see Adams-Bashforth) and the corrector (see Adam-Moulton), we can calculate the values of the local error truncation, which leads to the calculation of the optimal step-size for each of the g steps for the local group, and the global step-size can change by synchronizing the groups after m steps calculation, where m is typically larger than g.

A sample implementation of the explained algorithm based on Equations (7) and (14) can be described as follows:

1. Define the number of CPUs in group (g) based on current hardware restrictions.
2. Define both work group step size and work item step size.
3. Solve the starting points by using for example "DOPRI 5" and save them in $X_1, X_1, \dots, X_g$, and their corresponding derivations as $F_1, F_2, \dots, F_g$.
4. Update $X_{i+1}, X_{i+2}, \dots, X_{i+g}$ in parallel through the following steps:
    a. Calculate the derivation F by using Equation (7), Equation (14) and save in F_{i+1}.
    b. Wait for all values of $F_{i+1}, \dots, F_{i+g}$.
    c. Update the $X_{i+1}, X_{i+2}, \dots, X_{i+g}$.

   d. Calculated the error for each computing unit and then update the global step size.
   e. Synchronize all computing units and update value in global variable.

5. Go to the step 4 until all values calculated.

In previous, our implementation will be divided into 4 different parts:

(1) Calculating the gradient values for the next estimation points.
(2) Transferring the gradient vectors into the global memory.
(3) Estimating the next solution vector through an adapted Adams-Bashforth algorithm base on Equation (14) weights.
(4) Calculate local truncation error to adjust the step-size.

After these above listed main 4 steps, all local groups will be in synch (i.e., synchronized) for starting the next step. As we can see, most of the complexity of this algorithm is related to the correct usage of local and global variables, and most of the calculations will be solved in steps 3 and 4 depending on the problem size and the number of groups members.

An important effort in each calculation is to make the steps completely independent and create tasks with the same size in order to decrease the synchronization time between work groups.

Furthermore, by increasing the number of computing units to more than 32, we found out that this method then becomes inaccurate. For increasing the accuracy, the previous method is combined with an algorithm which is explained in Section 2.2, where the propagation factors G and F are replaced by a new suggested propagator factor. Therefore, the explained previous algorithm will run on local groups of processing units and grouped together does create a Parareal solver.

4. System Architecture

This computing system is designed based on the OpenCL platform. OpenCL is a heterogeneous computing platform, which is a framework for writing programs that are executed across heterogeneous platforms consisting of CPUs, GPUs, and other processors.

OpenCL includes a language (based on C99) for writing kernels (functions that execute on OpenCL devices) and APIs that are used to define and then control the platforms. OpenCL supports parallel computing by using a data-based and task-based parallelism. OpenCL has been adopted by Intel, AMD, NVidia, and ARM. Academic researchers have investigated the possibility of automatically compiling OpenCL programs into application-specific processors running on field programmable gate arrays (FPGA) [42]. Also, commercial FPGA vendors are developing tools to translate OpenCL to run on their FPGA devices. This feature of OpenCL motivates us to use this platform for implementing ODE solvers. But it is not possible to use this platform directly for different programming languages and web applications. Therefore, for the sake of expandability of the system, one application cloud, as shown in Figure 4, has been designed. This cloud application is creating a computing platform/network for solving ODE problems across a network of computing units.

Figure 3 is showing our global system architecture. It is composed of 3 main components. The ODE Computing platform OpenCL (we call it ODECCL) connectors are responsible for connecting the manager to the interfaces and getting/collecting tasks originating from different applications and destined to ODECCL.

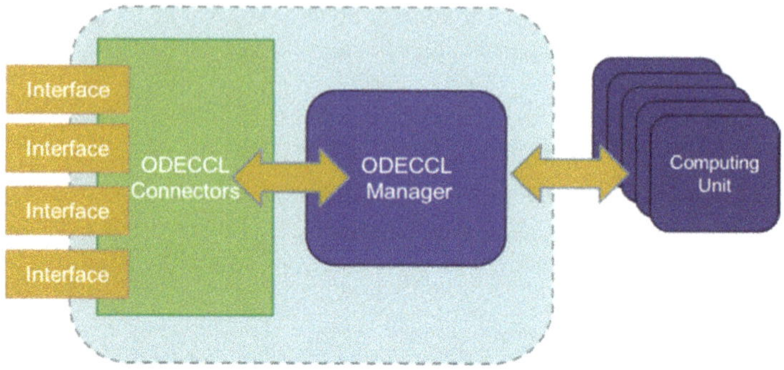

Figure 3. Solver system architecture. The interfaces provide the possibility of define specific tasks for the ODE solver. The ODE Computing platform OpenCL (we call it ODECCL) is composed of three parts. The manager will be assigning/allocating resources depending on their availability. Tasks are defined based on the different interfaces of ODECCL.

After a task has been validated in the system, it will be sending message(s) to the ODECCL manager. The ODECCL Manager is responsible for managing, scheduling and supervising tasks. Each task is scheduled based on its respectively needed computing unit's calculation power (Flops) and the communication cost. Computing units have the responsibility to execute tasks on the available OpenCL resources; therefore, it is possible to execute tasks both on CPU, n-CPU, and GPU at the same time (i.e., within the same parallel computing networked infrastructure).

Figure 4 does show the task scheduling concept in the ODECCL system. The scheduling is done between m groups and each group has g computing units (group units). After each stage, the computing units do exchange information in order to update their respective states and calculate the next gradient value.

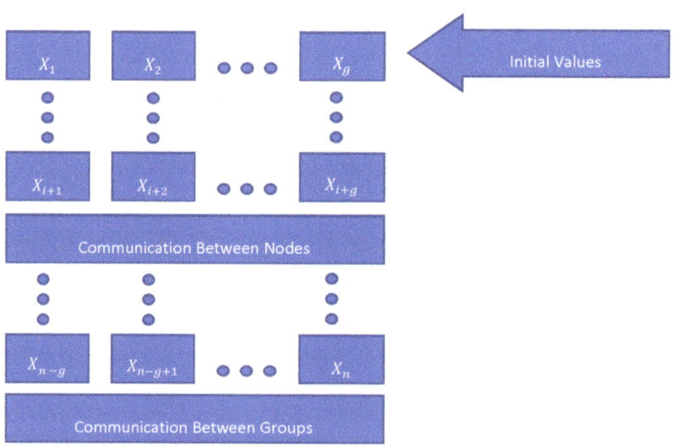

Figure 4. Scheduling scheme within/by the ODECCL system. Inside a group of processors, a stage will be processed, and between groups, steps will be calculated and synchronized.

From a technical perspective, ODECCL has been implemented using Visual Studio C++ and the OpenCL library provided by Nvidia has been included to the project. Each solver which is used in the experiments has been implemented as a kernel.

For example, if the use of the overshooting algorithm is not required, our system does use kernels without the overshooting parts. The manager part of the application is responsible to load the correct kernels for each problem and it is also responsible to create the groups and assign the required processing units to the created groups. The manager is also providing facility (i.e., infrastructure) to retrieve data from the different interfaces and load/download them from/to the kernels.

5. Numerical Experiments

For testing our computing system described in Section 4, we select three different types of ODEs. These three equations are shown in Equations (15)–(17). This does correspond to respectively solving Rayleigh (see Equation (15)), Rössler (see Equation (16)), and JACB (see Equation (17)) dynamic models.

The Rössler function is sensitive to inputs; see Equation (16). The parameters of the Rössler function are selected to have a chaotic behavior. This shall test our system in presence of small changes in initial conditions; thereby we can show significant variation in the observed behavior. This is also good to show the accuracy of the algorithm(s) while considering different computing units. Furthers, Equations (16) and (17) are selected as stiff ODE problems to test the system stability and for comparing our results with those from other related scientific works. They are very sensible to errors and a small error will/can change their respective final result.

$$\begin{array}{l} \frac{d^2x}{dt^2} - \varepsilon_1\left[1 - \left(\frac{dx}{dt}^2\right)\right]\left(\frac{dx}{dt}\right) + \omega x + k\sin(2\pi f_1) = 0 \\ \varepsilon_1 = 2.3\,,\,\omega = 1\,,\,k = 2.398\,,\,f_1 = 0.004 \\ X_0 = [-0.5, 0.1]\,,\,t \in [0, 20s] \end{array} \tag{15}$$

$$\begin{array}{l} \frac{dx}{dt} = -y - z \\ \frac{dy}{dt} = x + ay \\ \frac{dz}{dt} = b + z\,(x - c) \\ a = 0.2,\ b = 0.2,\ c = 5.7 \\ X_0 = [1, 1, 0],\ t \in [0, 20s] \end{array} \tag{16}$$

$$\begin{array}{l} \frac{dx}{dt} = y{\cdot}z \\ \frac{dx}{dt} = -x{\cdot}z \\ \frac{dz}{dt} = -0.51\,x.y \\ X_0 = [0, 1, 1],\ t \in [0, 7.5s] \end{array} \tag{17}$$

All models have been implemented on the following platform: Windows 10 PC with Intel Core i7 7700K as CPU, double Nvidia GeForce GTX 1080 TI with 8GB RAM as GPU and 64GB RAM. The Intel Core i7 7700K has 4 cores or 8 logical threads. The Nvidia GeForce GTX 1080 TI has 3584 cores which can be used for parallel computing.

Table 1 does show the respective kernel configuration for each of the solvers considered. As one can see, our model (PAMCL) does integrate two different solvers. The first solver has no overshooting algorithm and the second one has this ability of overshooting to fill up the problem of using a large number of cores. Regarding the first solver, the workgroup size is the same like the one of PIRK and PAM, and the number of working items is dependent on the number of available cores. But in the second solver of PAMCL the number of workgroups is variable and we always keep the number of internal working items of each workgroup to be 8, as this number is the most efficient solver w.r.t to the number of cores (as this is illustrated in Table 2).

Table 1. The kernel configuration for each of the solvers. * The PAMCL solver has two different types of kernels, the first one without overshooting and the second one with overshooting. If the number of cores is more than 32, the second kernel type is the one to be used.

Solver/Parameter	Work- Groups	Work Item	Kernel Type Number
PIRK	1	Depends on available cores	1
PAM	1	Depends on available cores	1
PAMCL	Variable	Depends on available cores divided by the number of Work-Groups	2 *

The computing time results of testing of our novel algorithm can be seen in Table 2, where they are provided for different differential equations to be solved. One can see by adding more cores results in increasing the performance of the system. But after 16 cores the performance increase is no more exponential, on one hand, and the overhead of the algorithm does significantly increase. Indeed, the decrease in computing time performance consecutive to increasing the number of cores 8 times, namely changing it from 8 cores and 64 cores, is just of 3 times, although one has added 8 time more cores. This poor gain in the resulting performance through adding more cores is even worser in when the number of cores is much higher.

Table 2. The execution time of PAMCL for different selected differential equations. The increase in number of cores does result in a decrease of the respective processing time. But by increasing the number of cores, this does result in more communication overhead amongst the cores.

Number of Cores on GPU	RAYLEIGH (ms)	JACB (ms)	RöSSLER (ms)
1	238.0	257.1	338
2	122.0	131.1	174.46
8	35.0	36.7	48.1
64	12.4	13.1	15.2
512	7.4	7.8	8.5

6. Comparison of the Novel Concept (PAMCL) with Previous/Related Methods

As explained in the previous sections, we define the speed-up by considering equal tasks on different cores. Therefore, our speed-up concept is not directly comparable to that of DOPRI or that of other algorithms, because most of them are rather running on a single-core computing unit. But for the sake of a better understanding, we calculate an "equivalent speed-up" performance w.r.t. one single computing unit. Figure 5 has been generated accordingly. As we can see, the PAMCL algorithm (i.e., our novel concept) can provide a much better speed-up depending on available free cores, either GPU or n-CPU. This speed-up, for 500 GPU cores, can reach up to 60x faster than the normal DOPRI5 algorithm, which is used in commercial ODE solvers like Matlab on the same computer/CPU.

In Figure 6, for our novel method PAMCL, the evolution of the processing time w.r.t. the CPU number is shown. As could be expected, according to the Gustafson's law [43], the system performance is not increasing linearly and the speed-up does reach a saturation after 16 computing units.

Further, in case of more complex equations, the advantage of our novel algorithm will increase, because the ratio between processing time and communication time to other processing units is higher.

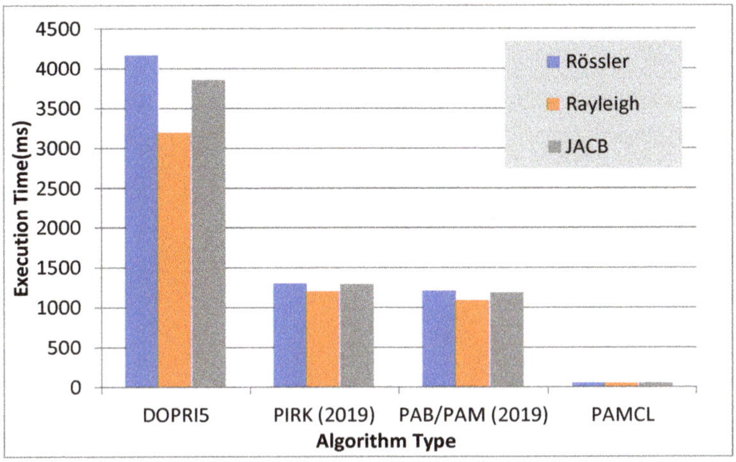

Figure 5. Comparison of the execution times of different problems while using different algorithms. The maximum computation error to stop the computing process is 0.001. All algorithms are executed on GPU for solving the Rössler equation. DOPRI5, PIRK, PAB/PAM and PAMCL are using 1, 8, 16 and 500 GPUs.

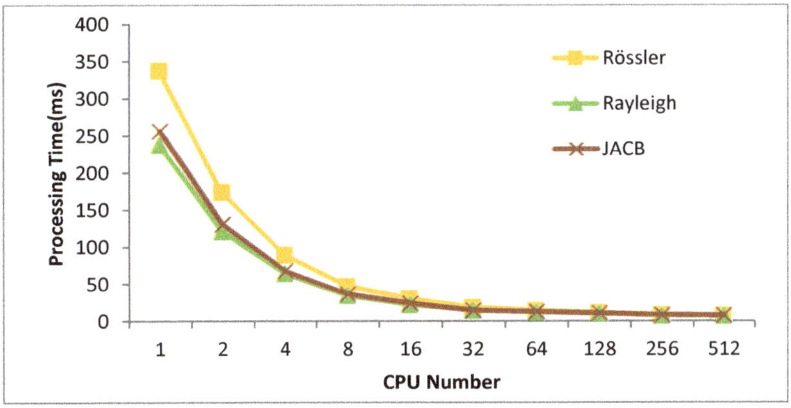

Figure 6. Effect of the number of processing units on the computing performance (for our novel approach PAMCL) for solving the Rössler attractor in GPU. The maximum error is 0.01.

While comparing CPU and GPU performance for solving differential equations, we do further observe that an implementation of the DOPRI algorithm on CPU is much faster than on GPU. However, while using the PAMCL algorithm, it does provide again more advantages w.r.t. a normal execution of the DOPRI algorithm on CPU (see Table 3).

For each model which has been explained previously, one has created a respective own kernel. The main problem regarding Runge-Kutta and PAM is that both methods have restrictions related to the number of cores as the number of cores increases beyond 8, both lastly named methods become worse w.r.t. reaching the required accuracy. Therefore, in order to reach the required error level, they will need more (i.e., additional) iterations, which does result logically in more computing time.

Table 3. Comparison of the "average computation times" on CPU and GPU while using different solver algorithms for solving the Rössler equation.

Method/ Algorithm	Error 0.01		Error 0.001		Error 0.0001	
	Time (ms) on CPU	Speedup (i.e., on GPU)	Time (ms) on CPU	Speedup (i.e., on GPU)	Time (ms) on CPU	Speedup (i.e., on GPU)
Dopri5	593.11	1x	4171.01	1x	41860.01	1x
PIRK (2019)	169.23	3.50x	1301.02	3.22x	15013.89	2.79
PAM/PAB (2019)	129.12	4.26x	1210.73	3.41x	11540.44	3.62
PAMCL	6.93	69.43x	51.83	80.47x	439.82	95.17x

As we can see in Table 4, the increasing performance which are demonstrated in previous table is due to the fact of using more global and local memory. Indeed, our model used respectively more memories compare to other methods.

Table 4. Comparison of "memory usage" w.r.t the target error while involving different solver algorithms for solving the Rössler equation.

Method/ Algorithm	Error 0.01		Error 0.001		Error 0.0001	
	Global Memory	Local Memory	Global Memory	Local Memory	Global Memory	Local Memory
Dopri5	18 KB	500 B	200 KB	500 B	2.1 MB	500 B
PIRK (2019)	19 KB	1.5 KB	210 KB	1.5 KB	2.2 MB	1.5 KB
PAM/PAB (2019)	22 KB	2.5 KB	250 KB	2.5 KB	2.6 MB	2.3 KB
PAMCL	32 KB	4 KB	350 KB	4 KB	4.8 MB	4 KB

As explained previously, for extending the PAMCL to a higher number of cores, we use the Parareal algorithm. This algorithm has its own drawback, as one needs to define the required iteration numbers needed to reach a convergence to the correct answer (see Figure 7). Thus, increasing the speed-up of PAMCL by using/integrating the Parareal algorithm will also have its own similar drawback.

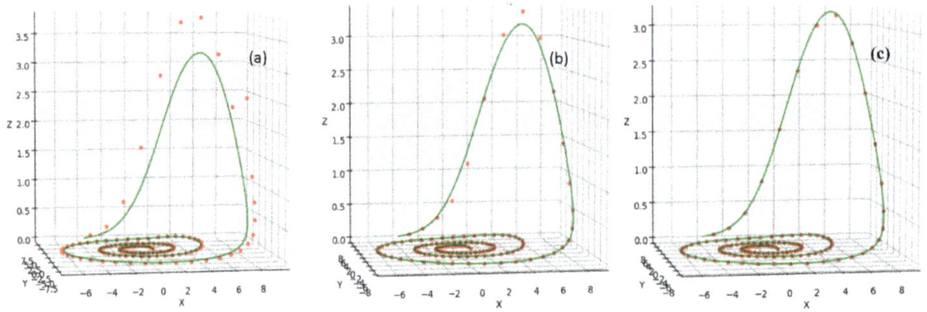

Figure 7. Showing the effect of iterations on the convergence towards the solution of the Rössler Equation (17). The green line is showing the expected solution, and the red dots are showing the PAMCL estimation at different iteration steps. For all 3 sub-figures (i.e., from left to right), the same parameter settings were used but 25 points are used in Sub-Figure (**a**), 50 points are used in Sub-Figure (**b**), and 100 points are used in Sub-Figure (**c**). Those points are calculated in a parallel way. It is visible that the estimation of model is changing from the expected results and an increasing number of does increase the model accuracy, but does also increase the calculation time.

7. Possible Extension of the PAMCL Model for also Solving PDE's

The suggested model (PAMCL) can also be used for solving PDE models. The main difference in solving PDE's lies essentially in the fact that in a PDE one has more dimensions. Therefore, it is possible to use/involve concepts such as Domain Decomposition, Waveform Relaxation [44] or multiple shooting method [45,46], which are used for solving either ODEs and PDE problems.

Regarding the Domain Decomposition approach, we can separate our domain into sub-domains. Then, in this case, we can solve each subdomain separately and combine the results of each domain to get the final solution.

As we have seen in previous sections, our model is not compatible with such a way of solving the problem and we use the multiple-shooting method for solving the ODE. Therefore, the best way to solve PDE problems in our model is to keep the domain and create subdomains in time (Multiple shooting method). This process can be expressed into the following steps:

1. Converting PDE problem into ODE problem. Customize solver to solve PDE in parallel on n-CPUs/GPUs groups.
2. Define the step-size of both coarse and fine estimators to find the solution of PDE between groups.
3. Solving the PDE sequentially using the coarse estimator in the overall time span.
4. Solving the PDE in parallel using fine estimator in each of the split time spans.
5. Update the values in each of the split time spans by using the coarse estimator sequentially.
6. Go to step 3 until we reach the required precision.

In step 1, we need to approximate/transform the PDE problem into an ODE problem. The approximated ODE problem now can be solved on our platform. This step normally requires setup parameters, calculates boundary conditions, and solves matrix solutions. Therefore, we need custom the solver for PDE solving. Each step of the PDE will be processed on group of CPUs\GPUs. In step 2, we a have similar implementation of PAMCL, we used both fine and coarse estimators to reach the final solution. First results are approximated, and later, by using the fine estimator they will be corrected. This process can be continued until the overall model reaches the required precision.

8. Conclusions

Using our novel PAMCL method for solving ODEs on an OpenCL framework does increase the performance. Compared to other solvers, our novel algorithm (PAMCL) is displaying very good behavior and does converge always to the exact solution. By choosing the correct interpolations and adjusting the weights, the algorithm can perform much faster calculations with the required precision. But still, the communication between computing units requires more optimization, and more unused resource do still exist in the system.

Solving these problems can provide much better performance w.r.t to the current status of the system. Also, as we see in the implementation part, defining equal tasks (by definition within the PAMCL algorithm) can increase the overall performance by decreasing task scheduling amongst nodes and does also increase the performance on a GPU like architecture, as an execution of branches on such structures are costly.

Also, by increasing the number of multi-stages, the calculation becomes more complex and it requires more resource to calculate values. This phenomenon occurs in all multi-step algorithms and it is required to provide load balancing between local workgroups and global workgroups.

Author Contributions: Conceptualization, V.T. and K.K.; Methodology, J.C.C. and K.K.; Software, V.T. and K.M.; Validation, V.T., K.M., J.C.C. and K.K.; Formal Analysis, V.T. and K.M.; Investigation, V.T.; Resources, V.T. and K.M.; Data Curation, V.T. and K.M.; Writing-Original Draft Preparation, V.T. and K.M.; Writing-Review & Editing, J.C.C. and K.K; Visualization, V.T. and K.M.; Supervision, J.C.C and K.K.; Project Administration, K.K. All authors have read and agreed to the published version of the manuscript.

Funding: This research received no external funding.

Conflicts of Interest: The authors declare no conflict of interest.

References

1. Sánchez-Garduño, F.; Pérez-Velázquez, J. Reactive-Diffusive-Advective Traveling Waves in a Family of Degenerate Nonlinear Equations. *Sci. World J.* **2016**, *2016*, 1–21. [CrossRef]
2. Neumeyer, T.; Engl, G.; Rentrop, P. Numerical benchmark for the charge cycle in a combustion engine. *Appl. Numer. Math.* **1995**, *18*, 293–305. [CrossRef]
3. Bajcinca, N.; Menarin, H.; Hofmann, S. Optimal control of multidimensional population balance systems for crystal shape manipulation. *IFAC Proc. Vol.* **2011**, *44*, 9842–9849. [CrossRef]
4. Baumgartner, H.; Homburg, C. Applications of structural equation modeling in marketing and consumer research: A review. *Int. J. Res. Mark.* **1996**, *13*, 139–161. [CrossRef]
5. Ilea, M.; Turnea, M.; Rotariu, M. Ordinary differential equations with applications in molecular biology. *Rev. medico-chirurgicala a Soc. de Medici si Nat. din Iasi* **2012**, *116*, 347–352.
6. Yadav, M.; Malhotra, P.; Vig, L.; Sriram, K.; Shroff, G. ODE—Augmented Training Improves Anomaly Detection in Sensor Data from Machines. In Proceedings of the NIPS 2015 Time Series Workshop, Montreal, QC, Canada, 11 December 2015.
7. Wang, X.; Li, C.; Song, D.-L.; Dean, R. A Nonlinear Circuit Analysis Technique for Time-Variant Inductor Systems. *Sensors* **2019**, *19*, 2321. [CrossRef] [PubMed]
8. Mahmoodi, S.N.; Jalili, N.; Daqaq, M.F. Modeling, Nonlinear Dynamics, and Identification of a Piezoelectrically Actuated Microcantilever Sensor. *IEEE/ASME Trans. Mechatron.* **2008**, *13*, 58–65. [CrossRef]
9. Omatu, S.; Soeda, T. Optimal Sensor Location in a Linear Distributed Parameter System. *IFAC Proc. Vol.* **1977**, *10*, 233–240. [CrossRef]
10. Pérez-Velázquez, J.; Hense, B.A. Differential Equations Models to Study Quorum Sensing. *Methods Mol. Biol.* **2018**, *1673*, 253 271.
11. Gander, M.J. Schawarz methods over the course of time. *Electron. Trans.* **2008**, *31*, 228–255.
12. Gander, M.J. The origins of the alternating Schwarz method. In *Domain Decomposition Methods in Science and Engineering XXI.*; Springer: Berlin/Heidelberg, Germany, 2014; pp. 487–495.
13. Niemeyer, K.E.; Sung, C.-J. GPU-Based Parallel Integration of Large Numbers of Independent ODE Systems. In *Numerical Computations with GPUs*; Springer: Berlin/Heidelberg, Germany, 2014; pp. 159–182.
14. Liang, S.; Zhang, J.; Liu, X.-Z.; Hu, X.-D.; Yuan, W. Domain decomposition based exponential time differencing method for fluid dynamics problems with smooth solutions. *Comput. Fluids* **2019**, *194*. [CrossRef]
15. Desai, A.; Khalil, M.; Pettit, C.; Poirel, D.; Sarkar, A. Scalable domain decomposition solvers for stochastic PDEs in high performance computing. *Comput. Methods Appl. Mech. Eng.* **2018**, *335*, 194–222. [CrossRef]
16. Van Der Houwen, P.; Sommeijer, B.; Van Der Veen, W. Parallel iteration across the steps of high-order Runge-Kutta methods for nonstiff initial value problems. *J. Comput. Appl. Math.* **1995**, *60*, 309–329. [CrossRef]
17. Seen, W.M.; Gobithaasan, R.U.; Miura, K.T. GPU acceleration of Runge Kutta-Fehlberg and its comparison with Dormand-Prince method. *AIP Conf. Proc.* **2014**, *1605*, 16–21.
18. Qin, Z.; Hou, Y. A GPU-Based Transient Stability Simulation Using Runge-Kutta Integration Algorithm. *Int. J. Smart Grid Clean Energy* **2013**, *2*, 32–39. [CrossRef]
19. Pazner, W.; Persson, P.-O. Stage-parallel fully implicit Runge–Kutta solvers for discontinuous Galerkin fluid simulations. *J. Comput. Phys.* **2017**, *335*, 700–717. [CrossRef]
20. Nievergelt, J. Parallel methods for intergrating ordinary differential equations. *Commun. ACM* **1964**, *7*, 731–733. [CrossRef]
21. Wu, S.-L.; Zhou, T. Parareal algorithms with local time-integrators for time fractional differential equations. *J. Comput. Phys.* **2018**, *358*, 135–149. [CrossRef]
22. Boonen, T.; Van Lent, J.; Vandewalle, S. An algebraic multigrid method for high order time-discretizations of the div-grad and the curl-curl equations. *Appl. Numer. Math.* **2009**, *59*, 507–521. [CrossRef]
23. Carraro, T.; Friedmann, E.; Gerecht, D. Coupling vs decoupling approaches for PDE/ODE systems modeling intercellular signaling. *J. Comput. Phys.* **2016**, *314*, 522–537. [CrossRef]
24. Bin Suleiman, M. Solving nonstiff higher order ODEs directly by the direct integration method. *Appl. Math. Comput.* **1989**, *33*, 197–219. [CrossRef]
25. Van Der Houwen, P.; Messina, E. Parallel Adams methods. *J. Comput. Appl. Math.* **1999**, *101*, 153–165. [CrossRef]

26. Godel, N.; Schomann, S.; Warburton, T.; Clemens, M. GPU Accelerated Adams–Bashforth Multirate Discontinuous Galerkin FEM Simulation of High-Frequency Electromagnetic Fields. *IEEE Trans. Magn.* **2010**, *46*, 2735–2738. [CrossRef]
27. Siow, C.; Koto, J.; Afrizal, E. Computational Fluid Dynamic Using Parallel Loop of Multi-Cores Processor. *Appl. Mech. Mater.* **2014**, *493*, 80–85. [CrossRef]
28. Plaszewski, P.; Banas, K.; Maciol, P. Higher order FEM numerical integration on GPUs with OpenCL. In Proceedings of the International Multiconference on Computer Science and Information Technology, Wisla, Poland, 18–20 October 2010.
29. Halver, R.; Homberg, W.; Sutmann, G. Benchmarking Molecular Dynamics with OpenCL on Many-Core Architectures. In *Parallel Processing and Applied Mathematics*; Springer: Cham, Switzerland, 2018.
30. Rodriguez, M.; Blesa, F.; Barrio, R. OpenCL parallel integration of ordinary differential equations: Applications in computational dynamics. *Comput. Phys. Commun.* **2015**, *192*, 228–236. [CrossRef]
31. Stone, C.P.; Davis, R.L. Techniques for Solving Stiff Chemical Kinetics on Graphical Processing Units. *J. Propuls. Power* **2013**, *29*, 764–773. [CrossRef]
32. Markesteijn, A.; Karabasov, S.; Glotov, V.; Goloviznin, V. A new non-linear two-time-level Central Leapfrog scheme in staggered conservation–flux variables for fluctuating hydrodynamics equations with GPU implementation. *Comput. Methods Appl. Mech. Eng.* **2014**, *281*, 29–53. [CrossRef]
33. Butcher, J. General linear methods. *Comput. Math. Appl.* **1996**, *13*, 105–112. [CrossRef]
34. Van Der Veen, W.; De Swart, J.; Van Der Houwen, P. Convergence aspects of step-parallel iteration of Runge-Kutta methods. *Appl. Numer. Math.* **1995**, *18*, 397–411. [CrossRef]
35. Fischer, M. Fast and parallel Runge–Kutta approximation of fractional evolution equations. *SIAM J. Sci. Comput.* **2019**, *41*, A927–A947. [CrossRef]
36. Fathoni, M.F.; Wuryandari, A.I. Comparison between Euler, Heun, Runge-Kutta and Adams-Bashforth-Moulton integration methods in the particle dynamic simulation. In Proceedings of the 4th International Conference on Interactive Digital Media (ICIDM), Bandung, Indonesia, 1–5 December 2015.
37. Bonchiş, C.; Kaslik, E.; Roşu, F. HPC optimal parallel communication algorithm for the simulation of fractional-order systems. *J. Supercomput.* **2019**, *75*, 1014–1025. [CrossRef]
38. Saha, P.; Stadel, J.; Tremaine, S. A parallel intergration method for solar system dynamics. *Astron. J.* **1997**, *114*, 409–415. [CrossRef]
39. Bellen, A.; Zennaro, M. Parallel algorithms for intial-value problems for difference and differential equations. *J. Comput. Appl. Math.* **1989**, *25*, 341–350. [CrossRef]
40. Lions, J.; Maday, Y.; Turinici, G. A parareal in time descretization of PDEs. *CR. Acad. Sci. Paris* **2001**, *I*, 661–668. [CrossRef]
41. Cong, N.H. Continuous variable stepsize explicit pseudo two-step RK methods. *J. Comput. Appl. Math.* **1999**, *101*, 105–116. [CrossRef]
42. Jaaskelainen, P.O.; De La Lama, C.S.; Huerta, P.; Takala, J.H. OpenCL-based Design Methodology for application-specific processors. In Proceedings of the 2010 International Conference on Embedded Computer Systems: Architectures, Modeling and Simulation, Samos, Greece, 19–22 July 2010.
43. Gustafson, J.L. Reevaluating Amdahl's law. *Commun. ACM* **1988**, *31*, 532–533. [CrossRef]
44. Gander, M.J. 50 years of time parallel time integration. In *Multiple Shooting and Time Domain Decomposition Methods*; Springer: Berlin/Heidelberg, Germany, 2015.
45. Wu, S.-L. A second-order parareal algorithm for fractional PDEs. *J. Comput. Phys.* **2016**, *307*, 280–290. [CrossRef]
46. Pesch, H.J.; Bechmann, S.; Frey, M.; Rund, A.; Wurst, J.-E. Multiple Boundary-Value-Problem Formulation for PDE-constrained Optimal Control Problems with a Short History on Multiple Shooting for ODEs. 2013. Available online: https://eref.uni-bayreuth.de/4501 (accessed on 3 August 2020).

sensors

Article

Power-Oriented Monitoring of Clock Signals in FPGA Systems for Critical Application

Oleksandr Drozd [1,*], Grzegorz Nowakowski [2], Anatoliy Sachenko [3,4], Viktor Antoniuk [1], Volodymyr Kochan [4] and Myroslav Drozd [1]

1 Institute of Computer Systems, Odessa National Polytechnic University, 65044 Odessa, Ukraine; viktor.v.antoniuk@gmail.com (V.A.); myroslav.drozd@opu.ua (M.D.)
2 Faculty of Electrical and Computer Engineering, Cracow University of Technology, 31-155 Cracow, Poland; gnowakowski@pk.edu.pl
3 Faculty of Transport, Electrical Engineering and IT, Kazimierz Pulaski University of Technology and Humanities in Radom, 26-600 Radom, Poland; as@wunu.edu.ua
4 Research Institute for Intelligent Computer Systems, West Ukrainian National University, 46027 Ternopil, Ukraine; volodymyr.kochan@gmail.com
* Correspondence: drozd@ukr.net; Tel.: +38-067-6873-880

Abstract: This paper presents a power-oriented monitoring of clock signals that is designed to avoid synchronization failure in computer systems such as FPGAs. The proposed design reduces power consumption and increases the power-oriented checkability in FPGA systems. These advantages are due to improvements in the evaluation and measurement of corresponding energy parameters. Energy parameter orientation has proved to be a good solution for detecting a synchronization failure that blocks logic monitoring circuits. Key advantages lay in the possibility to detect a synchronization failure hidden in safety-related systems by using traditional online testing that is based on logical checkability. Two main types of power-oriented monitoring are considered: detecting a synchronization failure based on the consumption and the dissipation of power, which uses temperature and current consumption sensors, respectively. The experiments are performed on real FPGA systems with the controlled synchronization disconnection and the use of the computer-aided design (CAD) utility to estimate the decreasing values of the energy parameters. The results demonstrate the limited checkability of FPGA systems when using the thermal monitoring of clock signals and success in monitoring by the consumption current.

Keywords: safety-related system; component; FPGA-designing; logical and power-oriented checkability; hidden faults; clock signal; consumed and dissipated power; temperature and current consumption sensors

Citation: Drozd, O.; Nowakowski, G.; Sachenko, A.; Antoniuk, V.; Kochan, V.; Drozd, M. Power-Oriented Monitoring of Clock Signals in FPGA Systems for Critical Application. *Sensors* **2021**, *21*, 792. https://doi.org/10.3390/s21030792

Received: 29 November 2020
Accepted: 22 January 2021
Published: 25 January 2021

1. Introduction

It can be argued that the presence of analogies from the natural world has led to the development of the human-created computer world. In this regard, energy is a central aspect of all living (biological) systems. Energy is received from the sun, from volcanoes at the bottom of the ocean, where flora and fauna bloom profusely, and many other natural sources. Computer systems are also powered by energy sources. The thermometer allows us to detect the deviations from the normal thermo state in living systems but also in artificial ones. As such, there are well-known studies on the monitoring of digital circuits using thermal sensors [1–3].

In support of robust biological systems, energy balance must be maintained by imposing a coordinated ordering (synchronization) of vital life processes. The synchronization functions in computer systems act much the same way in maintaining component integrity; however, they are much simpler. As such, turning the synchronization circuits off disables the components of a computer system and disrupts its operation without turning off the power. Failures in synchronization of digital circuits lead not only to the creation of a terminal disabling functionality but also to hidden failures, which can block error-free control circuits.

The concept of risk is the primary metric used in evaluating safety-related systems. Risk is determined by the product of two factors: (1) the probability of an accident and (2) the cost of the losses that it can cause. Trends in high-risk facilities are constantly increasing the importance of the second factor. However, reducing the probability of accidents [4,5] will certainly reduce the cost of losses but also result in a more robust safety-related system.

This task falls entirely on information technologies implemented in computer systems, which in safety-related applications are transformed into Safety-Related Systems, for example, safety systems of nuclear power plants [6]. According to international standards, these systems are aimed at solving the complex problem of functional safety: both the safety of the system and the facility under control in order to prevent accidents and reduce losses in case of their occurrence [7,8].

Computer systems perform many functions in various fields of production and consumption, but if they have become as significant in these fields as in accident prevention, for example, in healthcare, communications, or financial domains, it will follow that these fields have also become safety-related applications.

Safety-related applications have their own special significant impact on the synchronization problem. Safety-related systems are designed to operate in both normal and emergency modes. However, in emergency mode, the systems face the problem of insufficient checkability of their components, which causes the problem of hidden faults [9,10]. They can be covertly accumulated in normal mode, including synchronization circuits, and can create a real danger for emergency mode, collapsing fault tolerance of circuitry solutions and functionality of safety-related systems and facilities [11,12].

This paper describes the synchronization problems that take into account hidden faults inherent in safety-related systems and also presents a power-oriented monitoring of clock signals that is designed to avoid synchronization failure in computer systems such as a field-programmable gate array (FPGA). The former is based on the feature of synchronization, manifested in reducing the dynamic component of power consumption and by demonstrating how the power-oriented checkability increases in FPGA systems due to improvements in the evaluation and measurement of their energy parameters.

The main contributions consist in the following: (i) implementation of a power-oriented approach for monitoring the synchronization circuits in safety-related systems to counteract hidden synchronization outages in FPGA components, (ii) experiments conducted with the purpose of researching the thermal checkability of synchronization circuits in FPGA systems, (iii) experimental demonstration of the success in monitoring synchronization circuits by changing the consumption currents. Thus, the main challenge is related to improving the monitoring of synchronization circuits in FPGA components of safety-related systems. The key problem is focused on detecting a hidden shutdown of synchronization circuits in FPGA systems designed to operate in normal and emergency modes, based on a change in power parameters. The effectiveness of monitoring the synchronization circuits of FPGA systems by the energy parameters of the dissipated and consumed power is evaluated for the first time in this paper.

The rest of the paper has the following structure. Section 2 reviews the related works addressed to the development of thermal testability and thermal FPGA monitoring. In addition, the capabilities of modern computer-aided design (CAD) in the evaluation of energy parameters of FPGA systems are shown. Section 3 deals with the increasing problem of hidden faults, aggravating the consequences of synchronization failures, and related aspects of logical and power-oriented circuit checkability. The evolution of the traditionally used logical checkability is presented according to the resource approach. Due to the last one, the logical checkability is limited in the domain of safety-related applications, and there is a need to develop alternative forms, including power-oriented ones. Section 4 shows the results of experiments according to the evaluation of power-oriented checkability for FPGA systems and their monitoring to detect hidden faults in the synchronization circuits. The checkability of the circuits implemented in FPGA systems and the possibilities of their monitoring by the energy parameters of the dissipated and consumed power using temperature and current sensors are studied.

2. Related Works

Existing works related to the current problem can be divided into the following four groups: (i) thermal testability and thermal monitoring; (ii) capabilities of modern CAD in the evaluation of energy parameters in FPGA systems; (iii) monitoring of clock signals in FPGA; (iv) a need for developing the power-oriented checkability of circuits in safety-related applications, including monitoring of synchronization circuits in safety-related systems.

Within a *first* group, V. Székely et al. suggested the methodology of design for thermal testability " ... to find those chips that operate at higher temperatures than their normal operating temperature; to indicate possible mounting defects; to monitor during the whole lifetime in parallel with normal operation in order to warn of the danger of possible thermal runaway in advance ... " [13]. Yi Ren noted that "an all CMOS (complementary metal-oxide-semiconductor) temperature sensor test is proposed for submicron circuits to detect abnormal temperature changes so as to detect defects on a chip, and increase reliability and service life of devices" [14]. J. Altet and A. Rubio focused on the thermal testing of manufacturing catastrophic defects "to detect hot spots in the integrated circuits" [15].

An overview of the works on thermal testability and thermal monitoring of FPGA systems shows their focus on the overall assessment of thermal regimes. These studies do not analyze the ability to detect a particular type of fault, including synchronization failures.

Within a *second* group, modern CAD systems supporting FPGA-designing include utilities for the preliminary and current evaluation of energy parameters for developed schemes and offer external and internal sensors for their assessment. The utilities enable to set and account the activity of input and internal signals affecting the estimation of the dynamic component for the consumed and dissipated power of FPGA systems [16,17]. Electronics companies offer a wide range of external sensor chips that can monitor both chip temperature [18] and consumption currents for FPGA supply circuits [19]. In addition, some FPGA families have built-in crystal temperature monitoring and tools [20].

Monitoring of FPGA circuits on energy parameters is developed through support from CAD, which already provide utilities and sensors for assessing and measuring not only temperature but also currents characterizing power consumption. We can also note the improvement in sensor accuracy, which enables to enhance circuit monitoring in changing their energy parameters and creates conditions for fault detection with not only catastrophic changes in thermal modes for operation of integrated circuits.

Within a *third* group, C. Metra et al. noted in [21] that the checker of the self-checking register could not reveal a stuck-at fault affecting the clock signal and proposed the method for concurrently checking clock signal correctness in distribution networks of synchronous systems. This method and proposed "self-checking VLSI circuitry that concurrently checks clock signals for permanent and temporary faults which change signal waveforms from those expected from fault-free signals". Pei Luo and Yunsi Fei proposed to monitor the clock signals and detect glitches in FPGA for opposition to fault injection attacks in cryptographic applications by using a new scheme of comparison of the clock signals with reference clock [22]. G. L. Le et al. presented "a circuit and method herein for monitoring the status of a clock signal. The method includes supplying a pair of clock signals to a clock monitor circuit" for comparison [23].

Overall, we should notice that studies in the area of the clock signal monitoring in FPGA are dedicated to detecting clock glitches using reference clock signals and not the cause of these glitches. Indeed, supplying a clock signal to the register input does not guarantee that the register gets the clock signal because a circuit break can happen after the monitoring point. This loss of the clock signal can be detected by changing the power parameters.

The works of the *fourth* group are considered in the next separate Section 3. Because they not only note what has been done in the field under consideration but also determine the importance of developing energy-oriented approaches. Especially if we take into account new challenges associated with the development of safety-related systems and the limited capabilities of traditional logical checkability in relation to FPGA design.

3. Logical and Power-Oriented Checkability

Checkability is widely known in its simplest form—testability, i.e., the suitability of the digital circuit for the development of tests to identify its malfunctions. Testability is structural checkability as it is completely determined by the structure of the circuit [24,25].

During online testing, the checkability of digital circuits also depends on input data and becomes structurally functional. It is advisable to consider the development of circuit checkability in safety-related systems according to the resource approach [26].

Due to this approach, the integration of models, methods and tools combined by the concept of "resources" into the natural world are analyzed, and three levels in the development of resources are determined: replication, diversification and self-sufficiency as the goal of development. Replication is represented in the natural world by integration, which occurs due to a higher birth rate compared with mortality, which is typical of rodents, insects and bacteria. Replication is the simplest form of integration and will always be chosen in the absence of stamping restrictions, i.e., when there are open resource niches: market, technological, environmental and others. At this level, successful development is possible due to increasing productivity.

In today's computer world, replication is the dominant level of development. Hardware is stamped on the basis of matrix structures, including parallel shifters and adders, iterative array multipliers and dividers [27,28].

Large-sized software modules are stamped and connected to new software products to implement only a small amount of their functions. This negative process of slagging programs is supported by resource niches that are open for the performance and memory capacity of modern computers.

A niche filling process leads to closing resource niches when stamped clones are doomed to extinction. They can only survive if they show their own peculiarities and become individuals, that is, by moving to a higher level of development—diversification. Such a process is observed in green technologies, for example, in mobile systems where memory capacity and performance are dependent on the battery charge limit [29,30].

Safety-related applications stimulate closing of resource niches and moving to the level of diversification, where integration is based on increased trustworthiness, i.e., adequacy to the natural world, including aspects of functional safety. Under these conditions, computer systems rise to the level of diversification and become safety-related systems, diversifying the operating mode by dividing it into normal and emergency ones. The diversification process inherits the input data of digital circuits and their structural and functional checkability, which depends on these data that differ in normal and emergency modes, leading to the problem of hidden faults. It is important to note that this problem is inherent only in safety-related systems. In conventional computers, hidden faults do not create problems as they remain hidden during the full operating mode [31].

Hidden faults can occur not only due to the problem with the fault tolerance function of circuits in a very important emergency mode. This issue is better known for unsuccessful attempts to detect hidden faults while simulating emergency conditions. Unauthorized activation of these modes due to human factors or because of a malfunction has repeatedly led to accidental consequences. The controlled switching on simulation modes is not less dangerous due to the shutdown of emergency protection, which led to the Chernobyl disaster [32,33].

As described above, checkability is logical since it ensures carrying out logical control based on the detection of errors in the calculation results. Logical control is most widely employed in digital circuits, testing them during the pauses and in online testing using actual data [34,35].

The drawback of logical checkability is that it is limited to digital component circuits of modern safety-related systems. These circuits are traditionally based on matrix structures for processing data in parallel codes, which is the main reason for the low structural and functional checkability, especially in conditions of a little change in input data during normal mode. The current dominance of matrix structures requires the development of the concept of circuit checkability by diversifying its forms, among which the power-oriented form seems to be the most promising.

Works on circuit checkability and monitoring in order to detect violation of the thermal mode in integrated circuits due to power dissipation by measuring temperature, including thermal testability studies for safety-critical applications [36], developing online thermal monitoring methods [37], as well as techniques for measuring the temperature of local sections of FPGA microcircuits [38] are known today.

The interest in monitoring circuits due to power dissipation is clarified by the availability of temperature sensors. However, their limited accuracy has become a significant obstacle in the development of this direction. That enables the evaluation of the performance of FPGA systems as a whole with a significant change in energy consumption. Under these conditions, the task of power-oriented fault detection in synchronization circuits was not set.

Assessment of the current state of this issue and possibilities of the circuit power-oriented monitoring in synchronization circuits in the context of improving FPGA designing and its distribution in safety-related applications require experimental studies.

4. The Results of Experimental Studies

4.1. Experimental Conditions

The power-oriented checkability of circuits in synchronization chains is based on a decrease in power consumption of the dynamic component due to a decrease in the number of clock signal switching. The circuit is suitable for monitoring changes in the energy parameter beyond its values that are possible with proper functioning. During the experiments, lower values of energy parameters are estimated and compared at the correct operation of the investigated circuit and values of energy parameters for this circuit at disconnection of synchronization in bits of its registers.

The experiments evaluated the checkability of the circuits and their monitoring capabilities according to the parameters of the dissipated and consumed power. In the case of dissipated power, a series of experiments proofed the sufficiency of estimates obtained for FPGA systems using CAD utilities. To analyze power consumption, experiments are performed on the evaluation board stand, which enables measuring the consumption current of the built-in target FPGA chip directly. The experiments used the same target chip and CAD in both cases.

The experiments are carried out on the example of n bit iterative array multipliers, taking into account the activity of the input signals of the circuit, which was set at the recommended level of 12.5% of the clock frequency. For internal signals, a vectorless estimation was carried out.

Iterative array multipliers contain two n bit operand registers and a $2n$ bit product register, which receives clock signals with a frequency of CLK = 115 MHz, which is maximum for $n = 64$. FPGA systems of the multipliers are based on intellectual property (IP) Core LPM_MULT from the Library of parameterized modules [39], and they are implemented on the FPGA Intel Cyclone 10 LP: 10CL025YU256I7G target microcircuit under control of CAD Intel Quartus Prime 20.1 Lite Edition [40,41].

The selected FPGA contains 132 built-in 9 bit multipliers, on which LPM_MULT is implemented. The structure of the 9 bit multiplier (Figure 1) contains the input 9 bit busses of the operands *Data A* and *Data B*, reset *aclr* and synchronization *clock* inputs, as well as the output 18 bit bus of the result *Data Out*. The FPGA systems under investigation are based on built-in 9 bit multipliers.

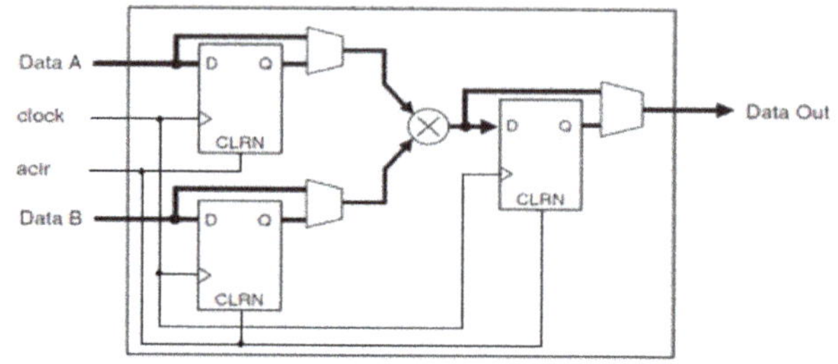

Figure 1. The structure of the built-in 9 bit multiplier.

4.2. Investigating the Thermal Checkability and Monitoring the Circuit of FPGA System

Modern CAD systems offer advanced tools for evaluating the energy parameters of FPGA systems, including the PowerPlay power analyzer utility, which enables estimating the dissipated power taking into account the dynamic and static components for the core, as well as the input/output system [41,42].

While evaluating power dissipation, the core is considered as all the structural elements of the FPGA chip. The power dissipation estimate is converted by the utility to the corresponding crystal temperature value.

Before performing calculations of energy parameters, it is necessary to set the predicted values of the input and internal signals activity of the FPGA system in the Power Play power analyzer. The activity or frequency of signal transitions in the system circuit has a decisive influence on the dynamic components of the consumed and dissipated power of the FPGA system.

For input signals, the activity can be estimated as a percentage of the value for the system clock signal or set by the number of transitions per second. The activity value of internal signals can also be set manually, similarly, to input signals, or calculated automatically by the utility based on the activity of inputs taking into account the structure of FPGA system circuit (vector less estimation).

In addition, before evaluating the utility, it is necessary to set the temperature and cooling system values for the system. They include defining the boundaries of the operating temperature range of the T_J crystal, as well as the ways of its determination.

The first option involves setting the supposed fixed value. The second one includes automatic calculation of the crystal temperature by the utility depending on the set parameters of the operating conditions, namely the ambient temperature and the selected cooling system. The lack or the availability of different types of chip cooling sets the corresponding values for thermal resistances between the crystal and the environment.

The results of investigating the FPGA system of the 32 bit iterative array multiplier are indicated in Table 1, including the power dissipation and the corresponding crystal temperature values with the activity of 12.5% for the input and internal signals of the circuit with synchronization disabling in d bits operand registers, where d = 0, 16, 32, 48.

Table 1. Dissipated power and crystal temperature of field-programmable gate array (FPGA) system when n = 32.

d	d,%	P_D, mW	P_{DCD}, mW	P_{DCS}, mW	P_{DIO}, mW	T_J, °C
0	0	166.16	18.61	73.25	74.30	30.0
16	12.5	163.96	17.06	73.25	73.65	29.9
32	25.0	157.17	11.33	73.22	72.62	29.7
48	37.5	153.05	7.99	73.21	71.85	29.6

The total power dissipation P_D is represented in milliwatts by its three components: dynamic P_{DCD}, static P_{DCS}, and dissipated power P_{DIO} of input/output system. These estimates are obtained employing the PowerPlay power analyzer.

Moreover, Table 1 shows the main change in power dissipation in its dynamic component. Synchronization disabling leads to a decrease in this component from 18.61 mW to 7.99 mW, which affects the total power dissipation in its decrease from 166.16 mW to 153.05 mW.

PowerPlay power analyzer calculates the crystal temperature:

$$T_J = P_D \cdot R_{JA} \cdot 10^{-3} + T_A, \quad (1)$$

where R_{JA}—crystal–environment thermal resistance, T_A—ambient temperature T_A = 25.0 °C.

R_{JA} thermal resistance is a constant for the FPGA system; its value depends on the presence or absence of a cooling system. In our experiment, R_{JA} = 30 is determined by CAD Intel Quartus Prime.

The temperature values T_J calculated according to Formula (1) for the corresponding values of the total power dissipation P_D are indicated in the right column of Table 1.

The circuit is suitable for monitoring synchronization failures in the event of a decrease in the energy parameter below the lower boundary of proper functioning. This boundary is determined by the utility with the smallest, i.e., zero activity of input and internal signals. Zero signal activity reduces the total power dissipation to the level of $P_{D.A0}$ = 148.83 mW. Then, the lower boundary of the dissipated power can be estimated, taking into account the utility error of ±2.5% [43] as $P_{D.MIN} = P_{D.A.0}$ (1 − 0.025) = 145.11 mW. In this case, the lower temperature boundary is determined by the Formula (1): $T_{J.MIN}$ = 0.14511 × 30 + 25 = 29.4 °C, where $P_D = P_{D.MIN}$.

Then, the checkability of the circuit makes it possible to detect synchronization failures when the crystal temperature drops below 29.4 °C.

While monitoring the circuit of the FPGA system in order to detect synchronization failures in reducing the crystal temperature, one should also take into account an error of the temperature sensors, the best of which introduce an error of 0.1 °C [44,45].

Therefore, monitoring can only detect failures that decrease the temperature below 29.3 °C.

The results of monitoring (Figure 2) show that when synchronization is disabled in 48 bits of registers (37.5% of the total number of circuit sync inputs), the temperature of the crystal drops from 30.0 °C to 29.6 °C and remains in the operating mode of the temperature values that the sensor can identify with the proper functioning of the circuit. Thus, the activity of the failures in the synchronization circuits is insufficient for its detection by monitoring the crystal temperature.

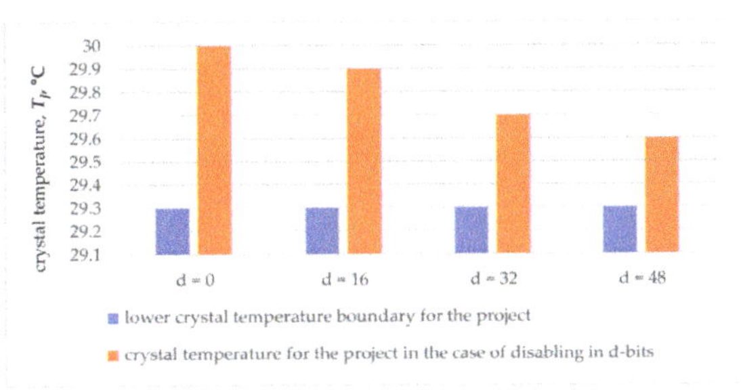

Figure 2. Results of thermal monitoring for 32 bit multiplier.

The results of investigating the FPGA system of the 64 bit iterative array multiplier are illustrated in Table 2, which also contains the values of dissipated power, its components and the corresponding crystal temperature. The last one is obtained using the PowerPlay power analyzer with the same signal activity of 12.5% and synchronization disabling in d bits of the operand registers when d = 0, 16, 32, . . . , 96.

Table 2. Dissipated power and temperature of FPGA system chip when n = 64.

d	d,%	P_D, mW	P_{DCD}, mW	P_{DCS}, mW	P_{DIO}, mW	T_J, °C
0	0	189.72	38.25	73.40	78.07	30.7
16	6.3	187.61	37.17	73.39	77.05	30.6
32	12.5	185.15	34.87	73.38	76.90	30.6
48	18.8	180.92	31.22	73.37	76.32	30.4
64	25.0	176.56	28.10	73.36	75.10	30.3
80	31.3	173.87	26.02	73.35	74.51	30.2
96	37.5	171.42	24.99	73.34	73.09	30.1

Table 2 shows the impact of synchronization disabling on the total power dissipation and its dynamic component by reducing disabling from 189.72 mW to 171.42 mW and from 38.25 mW to 24.99 mW, respectively. The crystal temperature calculated by PowerPlay power analyzer by Formula (1) when T_A = 25.0 °C is indicated in the right column of Table 2.

The lower boundary of the temperature $T_{J.MIN}$ of proper functioning is determined by the PowerPlay power analyzer due to the total power dissipation $P_{D.A0}$ = 160.90 mW, calculated when there is zero activity of the input and internal signals taking into account its decrease to the value $P_{D.MIN} = P_{D.A.0}$ (1 − 0.025) = 156.88 mW and with the utility error of ± 2.5%.

According to Formula (1), the lower temperature boundary is calculated as $T_{J.MIN}$ = 0.15688 × 30 + 25 = 29.7 °C, and it determines the suitability of the circuit of the FPGA system to control when the crystal temperature drops below 29.7 °C.

Monitoring of the FPGA system is carried out within the framework of the calculated checkability, which is reduced at the same time due to the error of temperature sensors.

The smallest error of 0.1 °C of the most accurate temperature sensors limits the monitoring capabilities in detecting failures to a temperature which is below 29.6 °C.

According to the results of monitoring (Figure 3), synchronization disabling in 96 bits of registers (37.5% of the total number of circuit sync inputs) reduces the temperature of the crystal from 30.7 °C to 30.1 °C, i.e., to a value that relates to the possible sensor readings with the proper functioning of FPGA system.

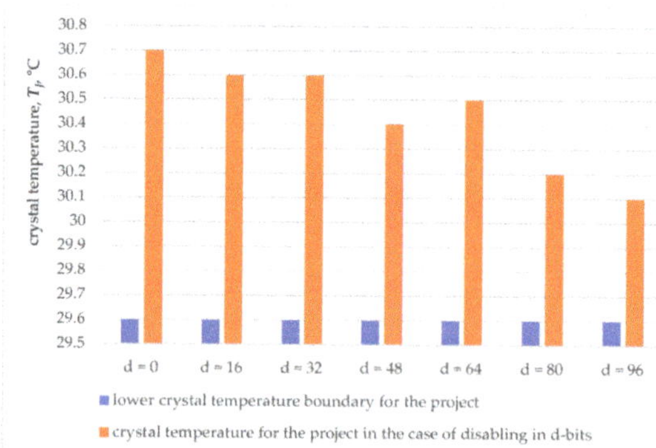

Figure 3. Results of thermal monitoring for 64 bit multiplier.

Thus, both considered examples indicate the impossibility of detecting failures in synchronization disabling even in 37.5% of the sync inputs of the circuit. Thermal monitoring is ineffective due to insufficient thermal checkability of the circuits in relation to synchronization failures and the limited accuracy of modern temperature sensors.

4.3. Investigating the Checkability of FPGA System and Its Circuit Monitoring by Current Consumption

Another possibility of power-oriented monitoring of FPGA systems is provided by the circuit checkability according to the parameter of power consumption, estimated by the current consumption [46].

The experiment was conducted using the evaluation board Intel Cyclone 10 LP FPGA evaluation kit [47], which enables to measure of the consumption current of the integrated FPGA chip. To display the measurement results, the evaluation board is connected to a personal computer.

The evaluation board contains the following components:

- Target Intel Cyclone 10 LP FPGA chip: 10CL025YU256I7G [40];
- Programmable generator of clock pulses programmable clock generator;
- Subsystem for measuring the current consumed by the target microcircuit;
- Toggle switches and push buttons to control the experiment progress.

Software. A formal description of the experimental device circuit, synthesis of the circuit, its placement and tracing in the space of the target FPGA chip was performed in the Intel Quartus Prime 20.1 Lite Edition CAD environment [41].

The target FPGA chip's consumption current is obtained and visualized in real time using the Power Monitor software utility [47], which is supplied by Intel along with the Intel Cyclone 10 LP FPGA evaluation kit. This utility is installed on a personal computer to display the ongoing consumption current values getting from the current measurement subsystem.

Description of the experimental scheme. The experimental scheme is complicated compared to the scheme for studying the dissipated power due to the need to generate input signals and implemented for n = 16, 32, 48 and 64.

For n = 16, the circuit (Figure 4) contains the phase-locked loop (PLL) module for generating the main clock signal, the OpUnit subcircuit for generating the multiplier operands, the subcircuit for disabling synchronization, consisting of 6 AND gates, and the investigated multiplier LPM_MULT.

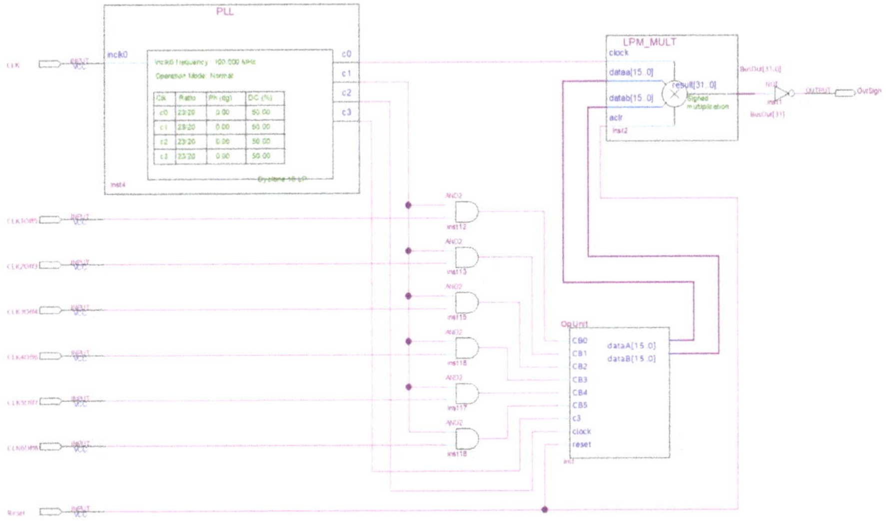

Figure 4. Experimental scheme for n = 16.

PLL module for generating the main clock signal. A clock signal with a frequency of 100 MHz is fed to the input of the circuit. This signal is formed by the programmable clock generator, which is part of the evaluation board. The phase-locked loop (PLL) module, which is part of the target FPGA chip, increases the clock frequency up to 115 MHz. This frequency is the maximum frequency for the experimental circuit. Four identical clock signals (115 MHz) are generated at the PLL module, and they are the main clock signals of the circuit.

OpUnit sub-circuit for forming the multiplier operands. Data for multiplier operands are generated in their registers composed of 4 bit *Johnson* counters ensured the uniform switching of all operand bits. The *Johnson* counter is implemented on the base of a cyclic shift register and controlled using a *clock* input and a shift *enable* input. The n bit operand registers contain $n/2$ *Johnson* counters, i.e., 8, 16, 24, and 32 for n = 16, 32, 48, and 64, respectively. Each *Johnson* counter generates a pair of bits for the multiplicand as well as a pair for the same multiplier bits. Each *Johnson* counter bit switches in 4 times less than the clock signal at the *clock* input. The clock signals are applied to *Johnson* counter clock inputs and the trigger that bisects the frequency and feeds the received signals to shift *enable* inputs of the *Johnson* counter. This reduces the switching frequency of the input signals (operand bits) by 8 times, i.e., decreases it to the recommended level of 12.5% of the clock frequency.

The OpUnit subcircuit for n = 16 is shown in Figure 5.

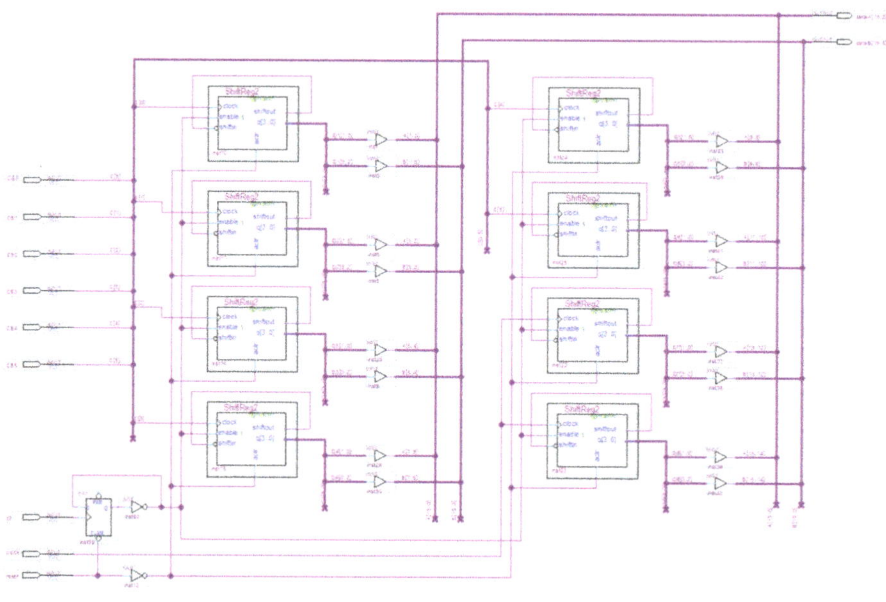

Figure 5. The OpUnit subcircuit for n = 16.

Subcircuit for disabling synchronization. To simulate a fault, the synchronization is disabled in six *Johnson* counters, which form the 12 least significant bits of both operands. To disable the synchronization, the 6 two-input AND gates and the 6 control signals are used, which are supplied to the circuit manually using both toggle *switches* and *push* buttons of the evaluation board to stand. These signals disable the synchronization of *Johnson* counters in bits (0, 1); (2, 3); (4, 5); (6, 7); (8, 9); (10, 11), respectively for both operands of the multiplier. This solution was sufficient for n = 16, 32, 48. A case n = 64 required more bits to disable the synchronization. For this, the circuit contains a shift register, which advances every 80 s the code disabling the synchronization of the next *Johnson* counter.

The multiplier is designed on the basis of standard IP Core LPM_MULT, similar to that used in experiments to estimate the dissipated power. It contains a multiplication circuit and an output register where the multiplication result is placed, as well as operand inputs, and the asynchronous reset input, and the clock input of the output register.

The experiment technique includes the step of preparing the evaluation board to stand and performing the measurements in operation three modes of the experimental scheme:

- Normal operation (mode 1);
- Operation at zero activity of information signals (mode 2);
- Operation under shutdown conditions of the operand register bits (mode 3).

During the preparation phase of the evaluation board stand, the experimental scheme is implemented into the target FPGA using the Intel Quartus Prime CAD by generating a configuration file with the following uploading in FPGA. The Power Monitor utility is activated to obtain and display the ongoing current consumption.

Further, the consumption current per each operating mode of the device was measured sequentially. Measurements per each mode are performed for one minute. The mode sequences are as follows:

- Normal operation;
- The zero-activity mode of the input signals, provided by stopping the formation of operands as well as a result of multiplication (the reset signal was supplied to the circuit and held for one minute);
- The mode of disconnecting the synchronization of *Johnson* counters, which form the bits 0..1, 0..3, 0..5, 0..7, 0..9, 0..11 and 0..11, 0..13, 0..15, 0..17, 0..19, 0..21, 0..22 of both operands for cases n = 16, 32, 48 and n = 64, respectively.

Results of experiments. The results of the studies are shown in Table 3 for n = 16, 32 and 48.

Table 3. Results of experiments for n = 16, 32 and 48.

d	d_{16},%	I_{16}, mA	d_{32},%	I_{32}, mA	d_{48},%	I_{48}, mA
0	0	15.74	0	20.99	0	29.47
4	6.3	13.72	3.1	18.97	2.1	27.85
8	12.5	12.51	6.3	16.95	4.2	26.24
12	18.8	10.49	9.4	15.34	6.3	24.62
16	25.0	8.88	12.5	13.32	8.3	22.20
20	31.3	6.86	15.6	11.30	10.4	20.18
24	37.5	5.25	18.8	8.88	12.5	18.16

Table 3 shows I_{16}, I_{32} and I_{48} values of consumption current in mA for n = 16, 32 and 48 at shutdown of synchronization in d bits. A case d = 0 corresponds to the normal mode. Experimental circuits contain the $4n$ synchronization inputs and determine the percentage of disconnections as $d_n = 100\ d/(4n)$: $d_{16} = 100\ d/64$, $d_{32} = 100\ d/128$, $d_{48} = 100\ d/256$. The zero-activity mode proofed the lowest values of 14.93 mA, 17.36 mA and 20.99 mA, determining the checkability of the circuits below these values for n = 16, 32 and 48, respectively. The error of 0.4 mA for the current sensor reduces the specified lowest values to the threshold values: S_{16} = 14.53 mA, S_{32} = 16.96 mA and S_{48} = 20.59 mA, a synchronization fault is detected below those threshold values. The results of experiments show the detection of a fault when the synchronization is disconnected in at least 4 (6.3%), 12 (9.4%) and 20 (10.4%) bits of operands for n = 16, 32 and n = 48, respectively.

The power monitor utility shows the current consumption graphs for n = 16 (Figure 6).

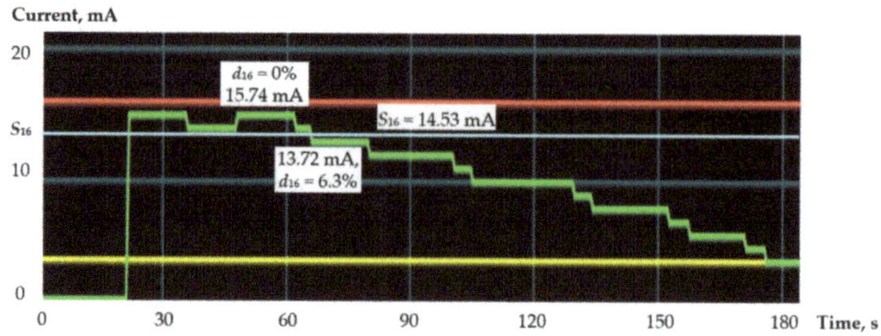

Figure 6. Current consumption charts for $n = 16$.

The red line marks the maximum consumption current value that corresponds to the normal operation of the device. The yellow line captures the minimum consumption value that corresponds to the last experimental mode (disabling the synchronization of bits 0.11 in both operands). The green line shows the current consumption values measured at each moment of the experiment. The charts show, from left to right, the level of normal mode (15.74 mA), zero activity of input signals, return to the normal mode, and the sequential sync shutdown mode. In the case of $d_{16} = 6.3\%$, the current consumption decreases to the level of 13.72 mA, which is below the threshold S_{16}. The results of the studies for $n = 64$ are shown in Table 4.

Table 4. Results of experiments for $n = 64$.

d	d_{64},%	I_{64}, mA	d	d_{64},%	I_{64}, mA
0	0	54.49	24	9.4	41.98
4	1.6	53.28	28	10.9	39.36
8	3.1	51.26	32	12.5	37.13
12	4.7	49.84	36	14.1	34.31
16	6.3	47.23	40	15.6	32.29
20	7.8	44.40	44	17.2	30.27

The zero-activity mode determined that the circuit checkability ($n = 64$) is less than 37.94 mA. The last one is reduced by the sensor error to a threshold of $S_{64} = 37.54$ mA, and the monitoring detects a synchronization fault below that threshold. Fault detection occurs when synchronization is disabled at least in 32 (12.5%) bits of operands.

Figure 7 shows the percentage of synchronization outages starting from which and above the monitoring provides the fault detection in circuit synchronization.

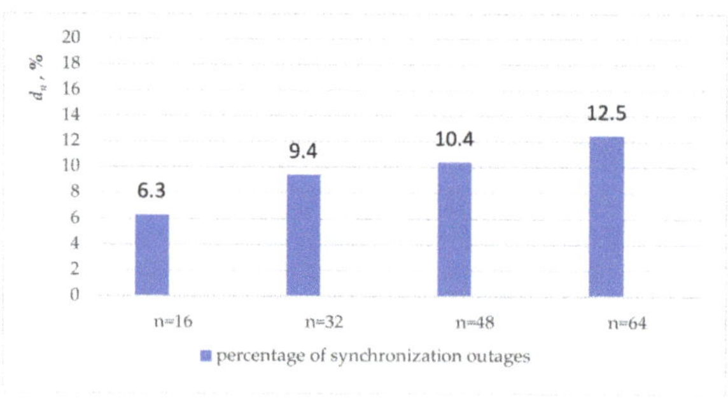

Figure 7. Comparison of results obtained in experiments.

For example, for $n = 16$ the sync fault is detected when 6.3% of the sync inputs are disconnected and with a higher percentage of disconnections.

The growing percentage of synchronization outages with a rise of n proofs a corresponding decrease in the detection ability of the monitoring. That can be explained by the nature of the dependencies on n the number of synchronization inputs and the complexity of the experimental scheme. The number of synchronization inputs grows linearly and increases by 4 times. The complexity of the circuit is mainly determined by the complexity of iterative array multipliers, which grows quadratically and increases by 16 times. Under these conditions, the switching of information signals increases its influence on power consumption to a greater extent compared with clock signals through both functional transitions and glitches [48–50].

The experiment was continued for the FIR7531 non-recursive filter, which is a low-pass filter with a cutoff frequency of 32.45 MHz and a gain of 16. Its implementation (fir_filter project) is included in the Intel Quartus Prime 20.1 Lite Edition CAD demo projects [41]. All basic FIR (finite impulse response) filter modules are described in Verilog. The FIR filter multiplier is implemented on the Intellectual Property Core LPM_MULT from the Library of parameterized modules (LPM_MULT library module). The FIR filter processes an 8 bit operand and forms an 8 bit result at the output using the two clock signals: clk and clkx2. The signal clk clocks the main circuit of the FIR filter, and the signal clkx2 clocks the result register.

The experimental circuit, implemented in the evaluation board, contains 32 subcircuits consisting of an FIR filter and an 8 bit *Johnson* counter to form an 8 bit operand, as well as a 3 bit counter that reduces the activity of input signals to the recommended level of 12.5%, and gates to disable sync signals. The zero-activity mode, provided by the reset of all *Johnson* counters, determined the testability of the circuit below 110.19 mA, reduced by the sensor error to a threshold value of S_F = 109.79 mA. Monitoring detects a synchronization failure below this threshold.

Table 5 shows the results of monitoring the FIR filters, including the number of d and percentage of d_F subcircuits with the disabled synchronization and the current consumption I_F measured by the sensor in the evaluation board.

Table 5. Results of experiments for FIR (finite impulse response) filters.

d	d_F,%	I_F, mA	d	d_F,%	I_F, mA
0	0	136.03	4	12.5	106.56
1	3.1	127.95	5	15.6	103.74
2	6.3	121.50	6	18.8	96.07
3	9.4	111.81			

The current consumption graphs provided by the Power Monitor utility are shown in Figure 8.

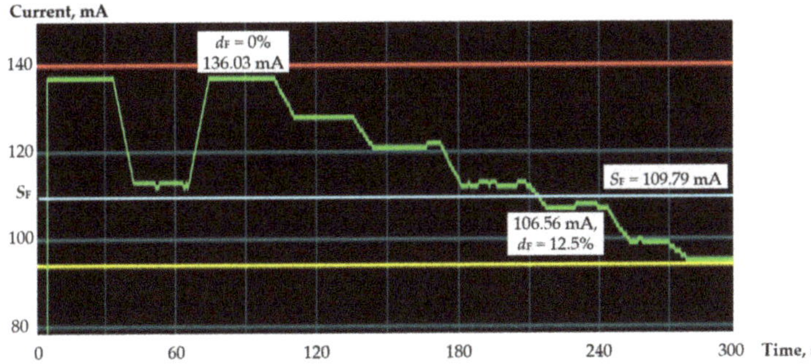

Figure 8. Current consumption charts for FIR filters.

The fault is detected when the synchronization is disabled in $d = 4$ ($d_F = 12.5\%$) and more subcircuits. The current consumption is reduced to 109.05 mA < S_F.

5. Conclusions

The problem of synchronization failures in digital circuits is brought to the attention because of their significant impact on functioning and possible blocking of logical control circuits aimed at detecting failures. This issue is aggravated in safety-related systems, which cannot only increase the price of failures in the conditions of prevention and elimination of accidents but also provide additional conditions for the manifestation of these failures in emergencies.

Experimental studies of power-oriented forms of checkability and monitoring capabilities of circuits for detecting failures in synchronization circuits are carried out using the iterative array multiplier implemented on an FPGA system under the control of CAD Quartus Prime. The energy parameters of the consumed and dissipated power are estimated due to the values of the consumed current of the FPGA system core and the temperature of its crystal, respectively. Failures in the synchronization circuits are introduced by disabling the sync inputs in the bits of the input registers of the iterative array multiplier.

Experiments conducted for studying the FPGA system due to power dissipation are carried out using the CAD utility and showed minor changes in temperature when the synchronization was disabled in 37.5% of the input register bits. This temperature change is completely blocked by the temperature operating range, taking into account the error of the temperature sensors and therefore excludes the possibility of monitoring the FPGA systems in order to detect failures in the register synchronization circuits by measuring the temperature.

An experimental study of the FPGA system due to the consumed power was carried out by measuring the consumption currents at the evaluation board stand and showed the suitability of the iterative array multiplier circuit for monitoring starting with synchronization disabling for 6.3% of the bits of the input registers ($n = 16$). With an increase in operands to $n = 64$, this figure increases to 12.5%, which is associated with the increasing effect of switching information signals (including glitches) compared to clock signals in the iterative array multiplier. An experiment with FIR filters showed the possibility of monitoring, starting with a 12.5% shutdown of sync signals.

Thus, the insufficient thermal checkability of circuits, further reduced by the limited accuracy of modern temperature sensors, does not allow monitoring the FPGA systems to detect failures in the synchronization circuits of registers due to power dissipation. However, this issue can be successfully solved by monitoring circuits due to current consumption, the measurement of which enables to detect of a synchronization failure, starting with 6.3% of disabled sync inputs.

Further research is planned in the direction of expanding the range of investigated schemes and studying the dependence of the results on the features of the initial data in accordance with the requests of customers related to the development of FPGA-based digital instrumentation and control safety systems for nuclear power plants.

Author Contributions: Conceptualization, O.D., G.N. and A.S.; formal analysis, O.D., G.N. and A.S.; investigation, G.N., V.A., V.K. and M.D.; methodology, G.N., V.A., V.K. and M.D. O.D. reviewed the problems of logical checkability of digital circuits and substantiated the need for the development of power-oriented forms of checkability. G.N. completed a review of thermal approaches to detecting failures and scheduled conducting the experiments. A.S. identified the problem of synchronization failures in FPGA systems and initiated investigating of this issue in FPGA designing. V.A. performed FPGA designing and conducted experimental studies taking into account the activity of input and internal signals. V.K. performed a review of temperature and current consumption sensors and participated in the experiments. M.D. combined synchronization problems with the problem of hidden faults and included the part dealing with safety-related applications of FPGA systems in the article. In addition, all authors participated in the evaluation of the results of the conducted experiments. All authors have read and agreed to the published version of the manuscript.

Funding: This research received no external funding.

Institutional Review Board Statement: Not applicable.

Informed Consent Statement: Not applicable.

Data Availability Statement: No new data were created or analyzed in this study. Data sharing is not applicable to this article.

Acknowledgments: The authors express their sincere gratitude to Robert Hiromoto, Professor of Computer Science, Center for Advanced Energy Studies 258, University of Idaho, USA, for his extensive editing of the English language of this paper.

Conflicts of Interest: The authors declare no conflict of interest. The funders had no role in the design of the study; in the collection, analyses, or interpretation of data; in the writing of the manuscript, or in the decision to publish the results.

References

1. Pauk, J.; Wasilewska, A.; Ihnatouski, M. Infrared Thermography Sensor for Disease Activity Detection in Rheumatoid Arthritis Patients. *Sensors* **2019**, *19*, 3444. [CrossRef] [PubMed]
2. Bartels, C.; Zhang, C.; Payá-Vayá, G.; Blume, H. A Synthesizable Temperature Sensor on FPGA Using DSP-Slices for Reduced Calibration Overhead and Improved Stability. In Proceedings of the International Conference on Architecture of Computing Systems, Porto, Portugal, 24–27 March 2015; Volume 9017, pp. 161–172. [CrossRef]
3. Enger, N. Care and Feeding of FPGA Power Supplies: A How and Why Guide to Success. *Analog. Dialogue* **2018**, *52*, 33–45.
4. Ivanchenko, O.; Kharchenko, V.; Moroz, B.; Kabak, L.; Konovalenko, S. Risk Assessment of Critical Energy Infrastructure Considering Physical and Cyber Assets: Methodology and Models. In Proceedings of the 10th IEEE International Conference on Intelligent Data Acquisition and Advanced Computing Systems: Technology and Applications, Lviv, Ukraine, 20–21 September 2018; pp. 225–228. [CrossRef]
5. Marvin, R. Risk Assessment. Common Cause Failures. Available online: https://www.ntnu.edu/documents/624876/1277591044/ccf.pdf/f435f724-469d-4492-860a-66eca10e6bd2 (accessed on 20 January 2020).
6. Kharchenko, V.; Sachenko, A.; Kochan, V.; Fesenko, H. Reliability and Survivability Models of Integrated Drone-Based Systems for Post Emergency Monitoring of NPPs. In Proceeding of the IEEE International Conference on Information and Digital Technologies, Rzeszov, Poland, 5–7 July 2016; pp. 127–132. [CrossRef]
7. IEC61508. *Functional Safety of Electrical, Electronic, Programmable Electronic Safety Related Systems. General Requirements*; Rep. IEC: Geneva, Switzerland, 2010.
8. IEC61513. *Nuclear Power Plants: Instrumentation & Control Systems Important to Safety. General Requirements for Systems*; Rep. IEC: Geneva, Switzerland, 2001.
9. Jung, J.; Ahmed, I. Development of FPGA-based reactor trip functions using systems engineering approach. *Nucl. Eng. Technol.* **2016**, *48*, 1047–1057. [CrossRef]
10. Smith, D.; Simpson, K. *The Safety Critical Systems Handbook*, 5th ed.; Butterworth-Heinemann: Oxford, UK, 2019.
11. Edstrom, J.; Tilevich, E. Reusable and Extensible Fault Tolerance for RESTful Applications. In Proceedings of the 11th IEEE International Conference on Trust, Security and Privacy in Computing and Communications, Liverpool, UK, 25–27 June 2012. [CrossRef]
12. Jhawar, R.; Piuri, V. Fault Tolerance and Resilience in Cloud Computing Environments. In *Computer and Information Security Handbook*; Morgan Kaufmann: Cambridge, MA, USA, 2017; pp. 165–181.
13. Ren, Y.; Wang, C.; Hong, H. An All CMOS Temperature Sensor for Thermal Monitoring of VLSI Circuits. In Proceedings of the IEEE Circuits and Systems International Conference on Testing and Diagnosis, Chengdu, China, 28–29 April 2009; pp. 1–5. [CrossRef]
14. Sarkany, Z.; Rencz, M. Design considerations to enhance thermal testability. In Proceedings of the IEEE Electronics Packaging Technology Conference, Singapore, 5–7 December 2012; pp. 148–152. [CrossRef]
15. Altet, J.; Rubio, A. *Thermal Testing of Integrated Circuits*; Springer Science + Business Media: Dordrecht, The Netherlands, 2002. [CrossRef]
16. Xilinx Power Estimator User Guide. Available online: https://www.xilinx.com/support/documentation/sw_manuals/xilinx2019_2/ug440-xilinx-power-estimator.pdf (accessed on 20 April 2019).
17. Intel Quartus Prime Standard Edition User Guide: Power Analysis and Optimization. 2018. Available online: https://www.intel.com/content/dam/www/programmable/us/en/pdfs/literature/ug/ug-qps-power.pdf (accessed on 20 March 2019).
18. Analog Devices Digital Temperature Sensors. Available online: https://www.analog.com/en/parametricsearch/11020 (accessed on 20 February 2020).
19. Analog Devices Power Monitors. Available online: https://www.analog.com/en/parametricsearch/11099 (accessed on 20 February 2020).

20. Intel FPGA Temperature Sensor IP Core User Guide. Available online: https://www.intel.com/content/dam/www/programmable/us/en/pdfs/literature/ug/ug_alttemp_sense.pdf (accessed on 20 February 2020).
21. Metra, C.; Favalli, M.; Ricco, B. Concurrent Checking of Clock Signal Correctness. *IEEE Design Test Comput.* **1998**, *15*, 42–48. [CrossRef]
22. Luo, P.; Fei, Y. Faulty Clock Detection for Crypto Circuits Against Differential Fault Analysis Attack. *IACR Cryptol. ePrint Arch.* **2016**, *967*, 1–8.
23. Li, G.M.; Richmond, G.J.; Raman, S. Circuit and Method for Monitoring the Status of a Clock Signal. U.S. Patent 7,454,645, 18 November 2008.
24. Chakrabarty, K.; Swaminathan, S. Built-in self-testing of high-performance circuits using twisted-ring counters. In Proceedings of the IEEE International Symposium on Circuits and Systems, Geneva, Switzerland, 28–31 May 2000. [CrossRef]
25. IEEE Std1500-2005. *Standard Testability Method for Embedded Core-Based IC*; IEEE: New York, NY, USA, 2005. [CrossRef]
26. Drozd, O.; Kharchenko, V.; Rucinski, A.; Kochanski, T.; Garbos, R.; Maevsky, D. Development of Models in Resilient Computing. In Proceedings of the IEEE Conference on Dependable Systems, Leeds, UK, 5–7 June 2019. [CrossRef]
27. Hiromoto, R. Parallelism and complexity of a small-world network model. *Int. J. Comput.* **2016**, *15*, 72–83.
28. Palagin, A.; Opanasenko, V. The implementation of extended arithmetics on FPGA-based structures. In Proceedings of the IDAACS International Conference, Bucharest, Romania, 21–23 September 2017. [CrossRef]
29. Murugesan, S.; Gangadharan, G. *Harnessing Green IT. Principles and Practices*; Wiley and Sons Ltd.: Hoboken, NJ, USA, 2012.
30. Tyurin, S.; Kamenskih, A. Green Logic: Models, Methods, Algorithms. In *Green IT Engineering: Concepts, Models, Complex Systems Architectures. Studies in Systems, Decision and Control*; Springer: Berlin, Germany, 2017; Volume 74, pp. 69–86. [CrossRef]
31. Drozd, A.; Antoshchuk, S.; Drozd, J.; Zashchoklin, K.; Drozd, M.; Kuznietsov, N.; Al-Dhabi, M.; Nikul, V. Checkable FPGA Design: Energy Consumption, Throughput and Trustworthiness. In *Green IT Engineering: Social, Business and Industrial Applications*; Springer: Berlin, Germany, 2019; Volume 171, pp. 74–94. [CrossRef]
32. Final Report on the August 14, 2003 Blackout in the United States and Canada: Causes and Recommendations. Available online: https://www3.epa.gov/region1/npdes/merrimackstation/pdfs/ar/AR-1165.pdf (accessed on 20 January 2020).
33. The True Toll of the Chernobyl Disaster. Available online: https://www.bbc.com/future/article/20190725-will-we-ever-know-chernobyls-true-death-toll (accessed on 20 January 2020).
34. Nicolaidis, M.; Zorian, Y.; Pradhan, D. On-Line Testing for VLSI. *J. Electron. Test. Theory Appl.* **1998**, *12*, 7–159. [CrossRef]
35. Metra, C.; Schiano, L.; Favalli, M.; Ricco, B. Self-Checking scheme for the on-line testing of power supply noise. In Proceedings of the Design, Automation and Test in Europe Conference, Paris, France, 4–8 March 2002; pp. 832–836.
36. Székely, V.; Márta, C.; Rencz, M.; Benedek, Z.; Courtois, B. Design for Thermal Testability (DFTT) and a CMOS Realization. In Proceedings of the 1st Therminic Workshop, Grenoble, France, 25–26 September 1995.
37. Székely, V.; Rencz, M.; Karam, J.M.; Lubaszewski, M.; Courtois, B. Thermal Monitoring of Self-Checking Systems. *J. Electron. Test.* **1998**, *12*, 81–92. [CrossRef]
38. Velusamy, S.; Huang, W.; Lach, J.; Stan, M.; Skadron, K. Monitoring Temperature in FPGA based SoCs. In Proceedings of the International conference on Computer Design: VLSI in Computers & Processors, San Jose, CA, USA, 2–5 October 2005; pp. 634–641.
39. Intel FPGA Integer Arithmetic IP Cores User Guide. Available online: https://www.intel.com/content/dam/www/programmable/us/en/pdfs/literature/ug/ug_lpm_alt_mfug.pdf (accessed on 20 February 2020).
40. Intel Cyclone 10 LP Core Fabric and General Purpose I/Os Handbook. Available online: https://www.intel.com/content/dam/www/programmable/us/en/pdfs/literature/hb/cyclone-10/c10lp-51003.pdf (accessed on 31 August 2020).
41. Intel Quartus Prime Standard Edition User Guide. Available online: https://www.intel.com/content/dam/altera-www/global/en_US/pdfs/literature/ug/ug-qps-getting-started.pdf (accessed on 31 August 2020).
42. PowerPlay Power Analysis. Quartus II Handbook Version 13.1.0. Available online: https://www.intel.com/content/dam/www/programmable/us/en/pdfs/literature/hb/qts/qts_qii53013.pdf (accessed on 20 April 2019).
43. Renbi, A.; Lindh, L. Power and energy efficiency evaluation for HW and SW implementation of nxn matrix multiplication on Altera FPGAs. In Proceedings of the 6th FPGAworld Conference, Stockholm, Sweden, 10 September 2009; pp. 45–51. [CrossRef]
44. Sachenko, A.; Kochan, V.; Turchenko, V.; Tymchyshyn, V.; Vasylkiv, N. Intelligent nodes for distributed sensor network. In Proceedings of the IEEE Instrumentation and Measurement Technology Conference, Venice, Italy, 24–26 May 1999; pp. 1479–1484.
45. Analog Devices LTC2986. Available online: https://www.analog.com/en/products/ltc2986.html#product-overview (accessed on 31 August 2020).
46. Drozd, O.; Antoniuk, V.; Drozd, M.; Karpinskyi, V.; Bykovyy, P. Power-consumption-oriented checkability for FPGA-based components of safety-related systems. *Int. J. Comput.* **2019**, *18*, 118–126.
47. Intel®Cyclone®10 LP FPGA Evaluation Kit User Guide. Available online: https://www.intel.com/content/dam/www/programmable/us/en/pdfs/literature/ug/ug-c10-lp-eval-kit.pdf (accessed on 31 August 2020).
48. Mangard, S.; Popp, T.; Gammel, B. Side-Channel Leakage of Masked CMOS Gates. In *CT-RSA 2005, LNCS*; Springer: Berlin/Heidelberg, Germany, 2005; Volume 3376, pp. 351–365. [CrossRef]
49. Drozd, A.; Drozd, J.; Antoshchuk, S.; Antonyuk, V.; Zashcholkin, K.; Drozd, M.; Titomir, O. Green Experiments with FPGA. In *Green IT Engineering: Components, Networks and Systems Implementation*; Springer: Berlin, Germany, 2017; Volume 105, pp. 219–239. [CrossRef]
50. Vikas, D. A review on glitch reduction techniques. *Int. J. Res. Eng. Technol.* **2014**, *3*, 145–148.

MDPI
St. Alban-Anlage 66
4052 Basel
Switzerland
Tel. +41 61 683 77 34
Fax +41 61 302 89 18
www.mdpi.com

Sensors Editorial Office
E-mail: sensors@mdpi.com
www.mdpi.com/journal/sensors

MDPI

www.ingramcontent.com/pod-product-compliance
Lightning Source LLC
Chambersburg PA
CBHW061553210926
51850CB00009B/2094